Concrete Handbook for Artists

Technical Notes for Small-scale Objects

Andrew Goss

Goss Design Studio

National Library of Canada Cataloguing in Publication Data

Goss, Andrew, 1944-
Concrete handbook for artists : technical notes for small-scale objects

Includes bibliographical references and index.
ISBN 0-9730223-0-2

1. Concrete sculpture--Technique. 2. Cement sculpture--Technique.
3. Ceramic jewelry. 4. Small sculpture. I. Title.

NB1215.G68 2002 731.4 C2002-900850-6

Cover design by Gordon Robertson
Cover art:
Small Arch #2: concrete, gold leaf, bronze, 4.5" X 8" X 2" (11 X 20 X 5 cm), 2000. A. Goss

Published by Goss Design Studio
781 2nd Ave West
Owen Sound, ON, N4K 4M2, Canada
www.makersgallery.com/concrete/handbook.html
519-371-1857 concrete@andrewgoss.com

Printed and bound in Canada
First edition. 1st printing April, 2002. 2nd printing and update April, 2006.

Contents

Preface
1 Definitions, 7
2 Technical: Concrete and Cement, 9
3 How-to: Introduction, 13
4 Hazards of Working in Concrete, 17
5 Materials: What to Use, 19
a Cement, 20
b Aggregates, 23
c Other Materials, 25
d Reinforcement, 29
e Colour and Finish, 32
f Premix, 35
6 Tools, 37
7 Mixes, 41
8 Setting (Hardening) and Curing, 49
9 Stresses and Strength, 53
10 Methods, 57
a Very Small Scale, 58
b Mesh: Step by Step, 65
c Casting: Simple Shapes, 75
d Surface Finishing, 83
11 Ciment Fondu (HAC), 87
12 Gallery, 89
13 Environmental Concerns, 101
14 Troubleshooting, 103
Appendix, 107
a Sources of Supply, 108
b Books, 112
c Links on the Internet, 114
d MSDS, 121
e Glossary, 127
Index, 131

Preface

I started working with concrete 14 years ago, primarily for aesthetic and philosophical reasons. I felt that the true value in the creative process lay in the beauty, art or craft of what we make with our hands and not in the cost of the material. Concrete was the ideal material, so I researched and experimented with it as a jewellery and small sculpture material.

Eight years ago I developed a web site, *www.makersgallery.com/concrete,* to share what I learned. As the web site grew in size it became shaped by what the many visitors wanted to know and wanted to share.

I realized the next step was to gather all the information into a printed form that would be easier to access. The book is now more than the web site but I hope it doesn't lose the connection. New information and materials can be added to the web site at any time to keep it up-to-date, and I hope readers will continue to visit the web site to contribute more information:
http://www.makersgallery.com/concrete/handbook.html.

In writing the book I have imagined the reader as a creative person, perhaps an artist, who would like to work with concrete and needs technical information. I have tried to write the book that I wished I had when I started working in this extraordinary material.

Most of the engineering, research and practical development in this area has been done in the construction industry. Suppliers of basic materials and additives are geared to that market as well. Hopefully I have made some connections between these two worlds.

Concrete is a vast subject. My intent is to explain the theory and principles of concrete, to give you the tools to understand what is happening, rather than the details in making a particular object. The book is about working directly with concrete, not about casting multiples. I have also intentionally set limitations of scale: jewellery, sculpture and small functional objects. It will not cover large outdoor works, such as cast-in-place sculpture.

Beyond the basic principles of concrete, the book explains mixes and additives and how these mixes must be changed for different uses. It explains several methods of forming concrete. Most of the techniques and mixes I have tried, and for those I haven't, I have given basic information from reputable sources along with leads for more research.

I knew when I started using concrete for jewellery and sculpture that I liked the material: I had used it in building one house and renovating several others, and I had used premixed concrete as the basis of some sculptures. But the more I worked with it the more I came to love it. When you first start working with concrete, it feels out of control. It won't stay where you want it, it's messy and dirty, it leaks out of the moulds. But then it sets and it becomes like stone – calm, solid, cool.

You can make it do almost anything. You can make a mould for it and the concrete will fill it. You can let it set for 12 hours, break the mould open carefully, and carve it. You can create a steel armature or structure and build up layers of concrete, each layer permanently bonding to the one underneath. It has extraordinary potential. I hope the information in this book sheds some light on that potential for you.

I would like to acknowledge many people who helped in the development of this book. Thanks to Stephen Hogbin, Bill Loney, Ilene Dunn, Rachel Reynolds, Joaquin Aviña and Nicholas Rees for allowing me to use images of their work, and for sharing their thoughts and expertise. Thanks also to Paul Tennis for checking the technical parts of the book, to Jane Cunningham for careful editing, to Gordon Robertson for a critical eye and to these friends for reading the manuscript: Rex Lingwood, John Newton and Bill Mason. I'm also grateful to Prime Gallery in Toronto and Harbinger Gallery in Waterloo who believed in concrete jewellery. And a thank you to the countless people who have emailed me over the last eight years with shared information, questions, books to read and additives to try. Creativity leads to generosity. And lastly I would like to thank my creative and generous wife, Sandra Noble Goss, not only for reading the manuscripts, but also for putting up with the dust, the mess, and the territorial intrusions in the studio.

Andrew Goss
March, 2002
Updated March, 2006

1

Definitions

It's important to get our terms straight because there is a lot of confusion between the words concrete and cement. Other relevant terms used in the book are defined in the **GLOSSARY**.

Concrete is defined as a mixture of crushed stone, sand, cement and water that hardens as a result of chemical reactions between the cement and water. Concrete means the wet mixture and also the hard material that results when it sets. (In English the word *concrete* comes to us via Old French, from the Latin *concretus*, which means 'grown together.')

Cement is generally defined as any substance used to make things stick together. In the context of this handbook, however, cement is Portland cement, defined as the fine grey powder that is made by burning limestone and clay. This is the active part of concrete that chemically combines with water to bind all the aggregates together. (*Cement* comes to us via the Old French *ciment*, from the Latin *caedmenta* meaning 'small broken stones.' Over time the word came to mean 'powdered stone' which is, in a sense, how we mean it today.)

Cement is not the same as concrete; cement is the binding part of concrete, cement is the active chemical. Concrete is the result of combining that cement with aggregates, water and other possible materials.

Cement paste is a mixture of cement and water. **Mortar** is a mixture of cement, sand and water. Mortar that is used for laying bricks or stones sometimes has added lime. **Concrete**, again, is a mixture of coarse aggregates, cement, sand and water. **Reinforced concrete** is concrete that has been strengthened by embedding a reinforcing material in it, usually steel.

This makes a logical progression of terms:

- Portland cement + water = **paste**
- paste + sand = **mortar**
- mortar + coarse aggregate = **concrete**
- concrete + reinforcing steel = **reinforced concrete**

Ferro-cement or **ferroconcrete** is a form of reinforced concrete using closely spaced multiple layers of steel mesh or small rods completely encased in a cement-sand mix. Ferro-cement is actually a misnomer. The technique was reinvented in the 1940s by Italian engineer Pier Luigi Nervi, who called it *ferro-cemento* and it came into English use as ferro-cement. A more accurate name might be ferro-mortar.

2

Technical: Concrete and Cement

Briefly...

Concrete is an artificially engineered material made from a mixture of Portland cement, aggregates (such as sand or gravel), water and other materials. Concrete is the most commonly used construction material in the world. It is strong, cheap and durable. Concrete's most important characteristic is that it is fluid and plastic when first mixed, but hard and extremely durable when set.

Portland cement combines with water (a chemical process called hydration) to bond the aggregates together into a solid whole. Portland cement is made from limestone or chalk, combined with silicates and a large amount of heat.

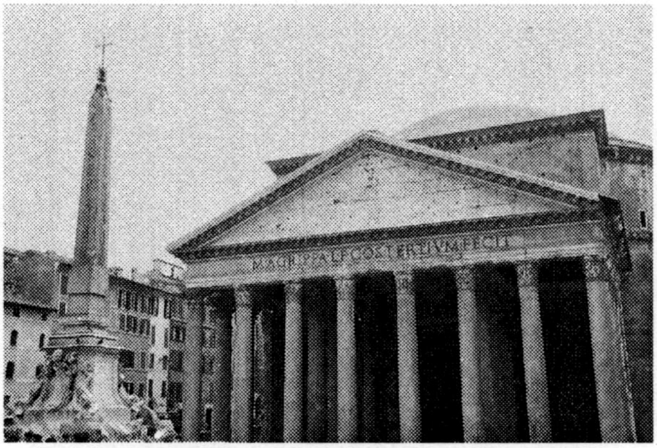

Pantheon: built in 27 B.C. in Rome using an early cement made with lime and volcanic pozzolan. [Arttoday]

Although cement-like materials were used by the ancient Egyptians and Romans – in 20 BC Vitruvius made cement from two parts pozzolan to one part lime – Portland cement as we know it today was patented by Joseph Aspdin, a builder in Leeds, England, in 1824. He mixed limestone (chalk) and clay and heated it on the kitchen stove, then ground it into a powder. He combined this powder with water and it set into a hard stone-like material. It was named after a stone it resembled that was quarried on the Isle of Portland. Portland stone (not Portland cement) was used in St. Paul's Cathedral in London, England, so the name "Portland" carried a lot of prestige.

We now use twice as much concrete in the world as the rest of the other building materials put together. The U.S. Geological Survey put the world's annual production of Portland cement at 1.7 billion metric tonnes in 2000.

Manufacturing Chemistry

Portland cement requires a source of calcium (usually limestone or chalk), silicates (sand, shale) and aluminates (clay, shale or sand). The raw materials are ground and mixed, then fired in a rotary cylindrical cement kiln up to 1,480°C (2,700°F), which partially melts the material.

Two reactions occur. First, the limestone or calcium carbonate turns into lime and carbon dioxide. Then the lime combines with the silicates to make dicalcium silicate (about 20%) and tricalcium silicate (about 55%) and with the aluminates to make tricalcium aluminate (10%) and tetracalcium aluminoferrite (about 8%). This is cooled and the resulting clinker is mixed with gypsum (about 3%) to control the setting speed, and ground to a fine powder called Portland cement.

Mixing Chemistry

When water is added to Portland cement and mixed, the silicates and aluminates combine with the water, rapidly at first, causing the initial set. The hardening of the cement binds the aggregate particles together into a solid whole. This process of hardening continues and slows down but never completely stops as long as there is moisture present. *Concrete does not harden because it dries; it hardens because it cures in the presence of moisture.* If concrete sets in one day and is kept continuously moist, it will be more than four times as strong after a week, six times as strong in a month, and more than eight times as strong after five years. These values are approximations, of course, and vary with mix design, temperature and other factors.

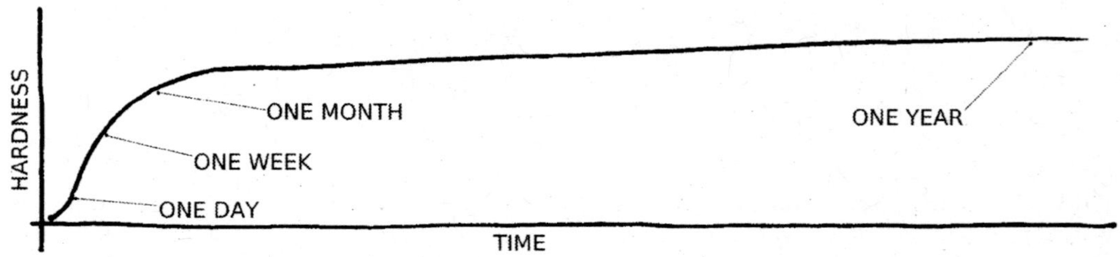

Diagram to illustrate how concrete sets over time.

There are many variables in the setting time for concrete (including temperature, water quantity and additives) so it's difficult to answer the question "how long does concrete take to set?" For more details see the **SETTING (HARDENING) AND CURING** chapter.

The two silicates mentioned earlier contribute most to the strength of Portland cement when they combine with water to make calcium silicate hydrates and calcium hydroxide. It is the calcium hydroxide (lime) that makes wet concrete alkaline, or caustic. The tricalcium aluminate also hydrates, but contributes primarily to early strength. The tetracalcium aluminoferrite contributes little to the cement except the grey colour. If cement is manufactured with low levels of iron and magnesium, white Portland cement is the result, with no loss of strength.

Before adding water, Portland cement is usually mixed with aggregates, typically sand and gravel, to make concrete for walls, floors, pillars, roadways or sidewalks. Ratios in the construction industry are typically 1:2:3 (cement:sand:gravel), 1:2:4 or 1:3:5. Mortar for laying bricks or blocks is usually a mixture of cement and sand with lime.

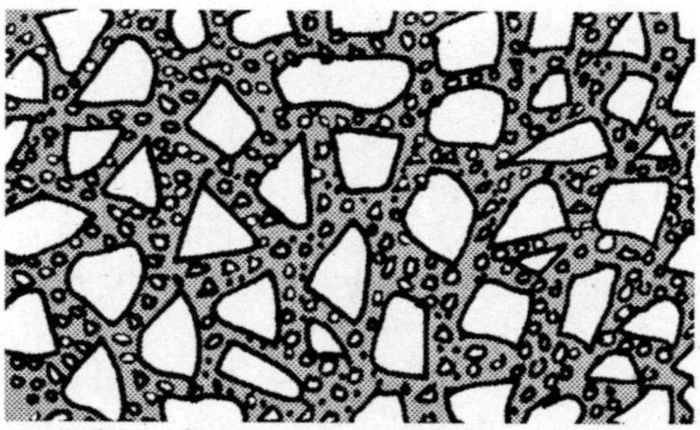

Concrete consists of large aggregate (gravel), fine aggregate (sand) and cement which holds the aggregates together.

The amount of water to be added to the mix is critical, knowing that the drier the mix the stronger the result. Water is usually one to one-and-a-half times the *volume* of the cement. One pound of cement can combine with only a little more than one third of a pound (0.38) of water. More water weakens the concrete. For example, after one month of curing, a mix of five gallons of water to a bag of cement, resulted in a strength of 5300 psi; nine gallons of water, 1600 psi. (Source: http://www.ag.ohio-state.edu/%7Eati_cons/256/Nbc_conc.htm)

There are many other possible additions to the mix:

- air-entraining chemicals for workability and frost resistance
- latex or acrylic for strength
- plastic fibres to control shrinkage cracks
- plasticizers (also known as superplasticizers or water-reducers) to permit less water and therefore create more strength
- steel reinforcing bars or mesh to reinforce it
- glass fibres (alkali resistant) for tensile strength
- pozzolans like silica fume or metakaolin to increase strength, decrease

permeability and improve shrinkage, by combining with the free lime
- pigments for colour
- accelerators or retarders to change the set time

What Happens Next

The water combines chemically with the Portland cement, which causes the cement to harden as it hydrates, holding the aggregates together along with any steel reinforcing. The process of hydration is exothermic; that is, it gives off heat. And the warmer the mixture gets, or the more heat you add, the faster the process becomes. (For very large structures – such as dams – this heat can result in such temperature differences between the inside and outside that cracking results.)

The concrete shrinks both as it sets and dries (.03% to .08%), sometimes producing shrinkage cracks that can become larger and more of a problem later on. High levels of shrinkage can be caused by excess amounts of water in the mix, high levels of cement, or conversely, low levels of aggregate. This is because the cement shrinks; the aggregate does not.

Premature drying, where the water evaporates from the surface before the concrete has set, also increases shrinkage. The concrete should be kept damp for as long as a month (one week minimum), the time when most of the curing/hardening takes place. Most shrinkage cracks occur in the first 24 hours, which is why polypropylene or glass fibres are sometimes added to the original mix. They prevent these cracks from starting. Because cement is hard, but has little tensile strength, it can also be brittle and chip easily. Latex or acrylic can be added (or substituted) for a portion of water in the mix to increase toughness, as well as contribute to watertightness. Both plastic fibres and latex are used in concrete canoe construction and have applications in artwork. See the **MATERIALS** chapter for more details.

Strength

Concrete has great strength in compression, but little tensile strength. Steel has great tensile strength, so the two are combined in bridges, buildings and other engineered projects. This is what is meant by "reinforced concrete." But even in small-scale artwork, steel wire and mesh can be used to reinforce the structure, particularly in unsupported spans. Alkali-resistant glass fibres are sometimes added for tensile strength, substituting for steel rebar. These are useful for thin applications where there is not enough thickness for steel. More details in the **STRESSES AND STRENGTH** chapter.

3

How-to Introduction

Briefly...

Uncured concrete is difficult to handle or shape directly. It usually needs some kind of form or structure to control the shape. This leads to three different fabrication methods:

1. casting into a mould – and later removing the mould
2. applying to an armature – expanded mesh, for example
3. carving a solid block – which has been partially cured

Concrete is a mix of Portland cement, an aggregate, water and other materials that give the mix certain characteristics. Portland cement is the fine, grey powder that binds the aggregates together.

Aggregates can vary depending on scale, colour and strength. The aggregates usually make up the bulk of the mix. Aggregates can be coarse like stone or gravel, or fine like sand. Other materials include pozzolans and admixtures.

General Rules

The dry materials are usually blended first. Clean water (or liquid, such as a latex or acrylic solution) is then slowly added to the dry mix while mixing thoroughly. Most measurements in this book are given as volumes, or relationships of volume. For more precise work, materials should be weighed, because air pockets in materials can lead to inaccurate volume measurements.

Important: **do not add too much water**; the drier the mix, the stronger the final product. Excess water that cannot chemically combine with free Portland cement stays in pores between the cured cement and weakens the structure. The pores can fill with water in the winter and freeze, cracking the concrete. How wet to make the mix depends on various factors but is often a function of the process: casting in a mould will need a wetter mix that slumps into place and fills crevices or undercuts. Applying the mix to a structural wire mesh will need a drier mix that feels more like clay and stays in place until it sets.

Cement will combine with only 38% of its weight in water. A common rule is: *one unit weight of cement to 0.40 unit weight of water*. But dry aggregate can take some of that water away; wet aggregate adds to the water, so measuring the water is not entirely accurate. I prefer to add water gradually until it's the "right" consistency.

Also, remember that wet concrete is thixotropic: a property of certain gels and mixes that become temporarily fluid when shaken or stirred, then revert back to the gel state. This means that when concrete is moved or vibrated it becomes more liquid. Contractors use this quality when they are pouring foundation walls, for example. They use vibrators immersed in the concrete (or applied to the outside of the formwork) to make the concrete more fluid so it flows around all the rebar and into corners. Caution: too much agitation will cause the coarse aggregate to settle to the bottom. This leads to different properties in different parts of the structure and can lead to strength problems or cracking.

After being placed, the mixed concrete is covered with plastic to prevent evaporation (or covered in wet burlap and misted with water) and left undisturbed for several days. Concrete will get somewhat hard after 4 to 12 hours, depending on the mix and temperature, and can be carefully removed from a mould or worked with coarse files and carved at this stage. In one or two days it becomes harder still, but it is possible to wet sand to refine the shape. After that it becomes comparable to working with stone.

If layering concrete, it is best to add layers of new concrete within the first day or two after the initial set, when almost complete bonding can take place. The surface should be damp, but not wet, before the next layer is applied. It is possible to add layers for several days after the basic structure is set, but bonding gets weaker. After about seven days the bonding will be only mechanical. Latex, or a cement-rich paste, can be used between layers at this stage to increase adhesion.

Finish

After final sanding and curing (in which the piece is kept covered and damp, or under water, for at least a week) the surface can be sealed. See **SURFACE FINISHING** for more details.

Safety

Safety issues will be expanded on in the **HAZARDS** chapter and in even greater detail in the **MATERIALS** sections, but here are some general guidelines to start. It is important to realize that cement-based materials are highly caustic (pH 10 to 12) and can cause damage to the skin (that may not be noticed for some time after contact), to the lungs, including long-term damage, and to the eyes. So:

- always wear gloves when mixing and working with concrete
- wear an appropriate mask when mixing concrete or sanding it
- protect your eyes

Read the **HAZARDS** chapter.

Storage

Cement

Cement stored in the open air will partially react with moisture over time and at the very least will reduce the strength of the concrete made from it. Seal open bags. I put mine in a plastic garbage bag, tied closed, and then put that in a large plastic garbage pail with a tight-fitting lid.

Sand

It doesn't matter if sand becomes damp, the important thing is that it stays clean. It can be stored outside, but keep leaves and other debris from falling in it by keeping it covered.

Other Materials

Keep dry materials off damp floors, in containers with lids, or in plastic bags tied shut. Liquids should be kept in jars with tight-fitting lids and proper labeling. All toxic materials should be kept safely out of the reach of children.

4

Hazards of Working in Concrete

There are serious hazards and health effects involved in some of these processes and materials and it is your responsibility to educate yourself about these hazards. This statement is not meant to dissuade you from working in concrete but it is a plea to be careful and inform yourself of the potential hazards. This chapter is only a starting point. Further details are available throughout the **MATERIALS** sections. Look for this symbol:

WARNING!

Please read the disclaimer at the beginning of this book.

Educate yourself about each of the materials you use by carefully reading the hazard warnings on the label, then follow that up with the MSDS (Material Safety Data Sheets). You can get these from the supplier where you purchased the product, or you can find them easily on the Internet searching for such terms as: 'MSDS Portland cement' or 'Material Safety Data Sheet' + 'fly ash'.

Concrete and Portland cement are **not** suitable materials for unsupervised children's art projects.

General Guidelines

Dust is a major problem in concrete projects. There is dust from cement, dry aggregates, both stone and sand, dust from other materials, admixtures and pigments, and most of these materials have hazards. Dust is thrown up from pouring, measuring and mixing these dry ingredients. Dust from cement can contain crystalline silica dust. Wear a mask, and make sure the mask (respirator) is approved for these dusts. If you are not sure, go to your local safety equipment and clothing supplier and explain to them what you need the respirator for. Make sure your mask is kept in a sealed bag or container when not in use. Wear waterproof gloves and long sleeves, as a lot of these dusts can be skin irritants. Wear eye protection as some of these dusts are eye irritants and can cause damage. Keep the workplace as free from dust as possible.

Store all materials and chemicals out of the way of children. Keep hazardous materials in non-food containers, and label them clearly.

When wet, Portland cement is extremely caustic and can burn your skin, so wear eye protection to prevent splatters, gloves and long sleeves to prevent skin contact. If you get it on your skin wash it immediately with soap and water. Use a barrier cream on your skin as an extra precaution before you put on protective gloves. Lack of pain is not an indicator that the cement is harmless. If you get it on your clothes, change them; otherwise the unset cement can continue to burn your skin. Third-degree burns are possible. More information in the **MATERIALS** sections and the MSDS in the **GLOSSARY**.

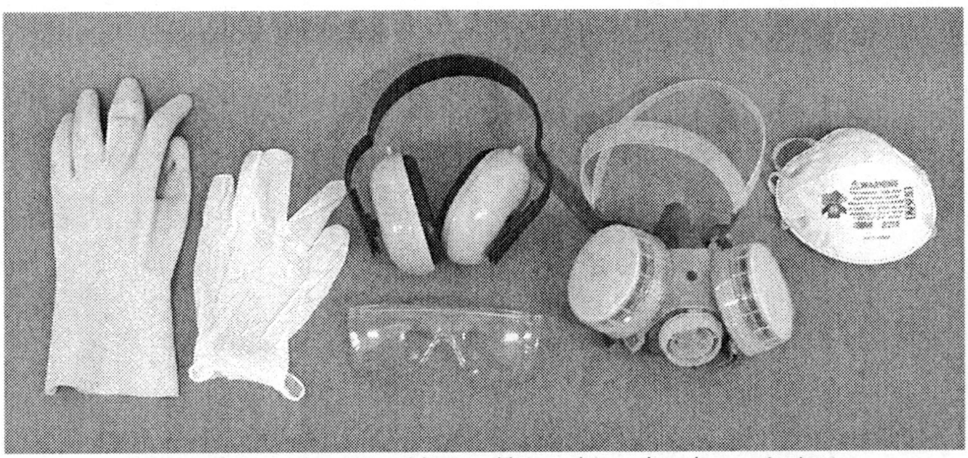

Some safety equipment: rubber and latex gloves, hearing protectors, safety glasses, toxic dust mask, nuisance dust mask.

For Further Reference

- ***Material Safety Data Sheets*** (MSDS) are available for any material that you buy that has safety issues.
- ***Artist Beware*** by Michael McCann, Lyons and Burford, New York, 1993. ISBN 1558211756.
- ***The Artist's Complete Health & Safety Guide*** by Monona Rossol, Allworth Press, 2001. ISBN 1880559188.

5

Materials: What to Use

a. Cement
b. Aggregates
c. Other Materials
d. Reinforcement
e. Colour and Finish
f. Premix

You can make good concrete with crushed stone, sand, cement and water, but if you want better strength, impermeability and workability, there are other materials you can add. The following sub-chapters describe these materials. For suppliers, check both the **SOURCES** and **LINKS** sections.

5a

Cement

Portland Cement

Portland cement (which is mostly calcium silicates) holds the aggregates together and is available in different grades and colours. It is usually sold in 94 lb. or 50 kg bags. The type you can buy at the local hardware or lumber store is grey in colour. Cement types are defined in the U.S.A. (and some other countries) by ASTM C150, Standards Specification for Portland Cement. In Canada, CSA A5 defines the requirements for Portland cements, and other countries generally have their own specifications. Cements from different sources, even of the same type, may behave slightly differently with respect to set time, strength gain, and interactions with admixtures or pozzolans. These are the U.S.A. ASTM standards, with Canadian CSA types in brackets from the Portland Cement Association, http://www.cement.org/tech/cct_cement_types.asp:

Type I (Type 10): is a normal, general-purpose cement suitable for all uses. It is used in general construction projects. For all intents and purposes this is the Portland cement you can locate easily and is the one to use.

Type II (Type 20): generates heat at a slower rate and has a moderate resistance to sulfate attack.

Type III (Type 30): is high-early-strength cement and causes concrete to set and gain strength rapidly. It is chemically and physically similar to Type I, except with finer particles. This may increase water demand and lower long-term strength.

Type IV (Type 40): has a low heat of hydration and develops strength at a slower rate than other cement types. It is rarely produced any more.

Type V (Type 50): is used only in concrete structures that will be exposed to severe sulfate action.

When an **A** appears after the type, e.g. "Type IA," the cement has an air-entraining agent already added. Some cements have two designations, e.g. Type I/II fits the requirements of both types.

White Portland cement has been manufactured largely free of iron and magnesium, resulting in a pure white powder, with no loss of strength compared to grey Portland cement. But it is more expensive and not as readily available. It is used where the colour has to be white or a pastel pigmented shade.

Blended Cements

These cements are a blend of Portland cement with other ingredients such as pozzolans (fly ash, silica fume, blast-furnace slag, see **5c OTHER MATERIALS**) for particular applications and they're slowly becoming more popular. By replacing Portland cement and using recycled materials, they are also more environmentally friendly. If you can find them they may also be more convenient than adding pozzolans yourself. As one example: Type IS has between 25% and 75% blast-furnace slag. Some of their different characteristics could include: lower heat, slower strength gain, higher ultimate strength, lower permeability, and better durability. Some may need longer curing to achieve these properties. ASTM C 595 specifications list the following types in the U.S.A.:

Type IS: Portland blast-furnace slag cement
Type IP and Type P: Portland-pozzolan cement
Type I(PM): Pozzolan-modified Portland cement
Type S: Slag cement
Type I(SM): Slag-modified Portland cement

Canada's CSA A362 standard allows five types of blended cement:

Type 10S: Portland blast-furnace slag cement
Type 10SM: slag-modified Portland cement
Type 10F: Portland fly ash cement
Type 10FM: fly ash-modified Portland cement
Type10SF: Portland silica fume cement

(Source: Portland Cement Association publication: PCA R&D Serial No. 2475, ISBN 0-89312-203-3)

WARNING!

Skin contact: Avoid all skin contact. Cement is caustic like lye (alkaline, pH 12 to 13) when mixed with water and not cured. Wet Portland cement will not only cause **skin irritation, it can also cause irreversible tissue damage, including third-degree burns**. Don't rely on pain or discomfort as an indicator at the time; several hours later the damage may become apparent.

As well, you may develop defatting dermatitis or **allergic dermatitis** because of chromium contaminants in the cement. Once you develop this sensitivity to cement, in the form of a skin allergy, you may not be able to use it any more. So use preventive measures to make sure you do not become sensitized. **Wear protective gloves** (a minimum of latex or vinyl – replace them when they are torn) and long sleeves. A barrier cream is not sufficient. If you get wet cement on your skin wash it immediately with soap and water.

Serious eye damage, ranging from irritation to burns and blindness: The chemical compounds in Portland cement are highly corrosive to the eyes. Wear safety glasses with side shields, or goggles.

Lungs: Airborne dry cement dust can cause immediate or delayed irritation or inflammation; wear a mask (respirator) rated for cement dust. Portland cement can have trace amounts of crystalline silica dust, which can lead to **silicosis**, a potentially fatal lung disease. If you get concrete dust in your clothes, clean them. Don't wear them around the house. Avoid methods of working which would add dust to the air. And don't eat or drink in the work area.

First aid: for eyes, flush with water for at least 15 minutes and seek medical attention. For skin: flush thoroughly with water. For inhalation: remove to fresh air. For ingestion: give victim water and seek medical attention.

Cement can also be a hazard to your plumbing. Do not wash your tools in the sink. The unset concrete may harden in your plumbing. Always wash tools in a separate container and let it settle for at least a day. Pour off the water and dispose of the concrete.

High Alumina Cement

High alumina cement (HAC) is often called ciment fondu. See the separate chapter on **CIMENT FONDU**.

WARNING!

Ciment fondu is considered a nuisance dust, and an irritant to eyes, skin, nose and throat. Wear a dust mask, eye protection and gloves.

5b

Aggregates

Aggregates are the substance and strength of most concrete mixes. A common saying in the concrete industry is that the concrete is only as strong as the aggregate.

Coarse Aggregates

Crushed stone or **gravel** can range in size from 1/4" to 1". For small-scale sculpture you probably won't need any stone larger than 1/4" or 5 mm. You may not need it at all unless cost is a factor: stone is cheaper than cement. Stone adds strength to larger work. It should be hard (not powdery or easy to break), and free of dirt or clay. Ask if it has been washed, if not rinse it clean.

Other coarse aggregates include **crushed brick** or **recycled concrete**, and other broken hard materials like **fired clay**. **Glass** is not advisable due to chemical reactions with some types of glass.

Fine Aggregates

Sand should be "sharp," called **sharp sand**, **brick sand** or **mortar sand**. The grains of pit-run sand or beach sand are more round, yielding a lower strength. The sand should be clean with no organic contamination; sand from salt-water beaches should be washed so no salt remains.

WARNING!

Wear a dust mask when handling or mixing aggregates. Concrete when set is like stone, so the same cautions that apply to stone carving and polishing apply. Wear an appropriate mask (one approved for toxic dust) and eye protection when sanding. Wet sand whenever possible to minimize dust. The dust from dry sanding can cause **lung problems**: **silicosis**, **emphysema**, **bronchitis**, **fibrosis**. Some aggregates in the concrete can, when sanded into dust, be dangerous, usually from the free silica. The worst aggregates for this are sandstone, granite, slate and quartz. If you are carving concrete wear eye protection.

Lightweight Aggregates

These materials are added for their insulating properties, to lighten the weight, or to make the concrete easier to carve. Note: these aggregates will weaken the concrete in proportion to the amount used.

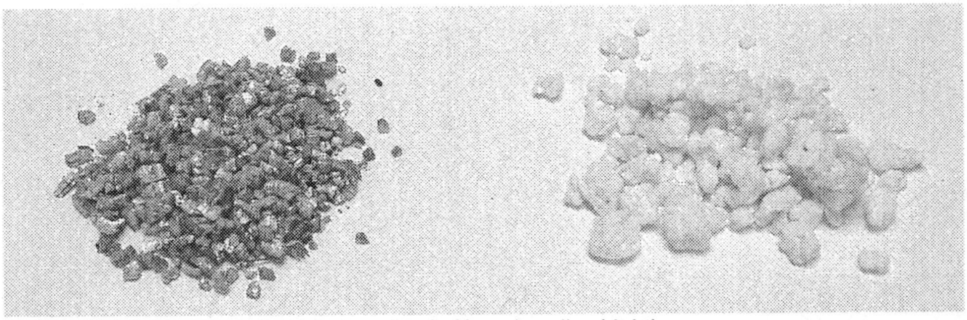

Vermiculite (left) and perlite (right).

Vermiculite: A naturally occurring, non-toxic, soft mineral that resembles mica. It is heated to a high temperature to produce a lightweight, sterile, laminated structure. Used for plant soils and as an insulating material.

Perlite: A naturally occurring volcanic glass, it expands when heated to a whitish-grey material used for insulation and as a planting material.

Pumice: A volcanic (glass) rock that is very light, full of air bubbles.

WARNING!

Some vermiculite, perlite and pumice may contain silica (**silicosis, lung cancer**), so wear an approved dust mask when working with it.

Glass spheres: These hollow glass balls, sometimes called 'macrospheres' are about the size of buckshot or slightly bigger.

Styrofoam® Beads or **polystyrene beads**: A lightweight plastic used for insulation and beanbag chairs. The beads are difficult to work with as they are resistant to wetting and float in the mix. Coated beads are available.

5c

Other Materials

Very Fine Materials

Stone dust, a waste product from quarries or stone works, can be added to smooth mixes for small-scale work. It adds strength, reduces shrinkage on setting, and may add to the visual appearance of the concrete. It also helps make the mix more plastic. **Limestone** and **marble dust** are two types of stone dust. (Cautions: stone dust may necessitate using more water and may weaken concrete in large-scale work. It is not a recommended material for construction purposes.)

WARNING!

Stone dusts can have the same health effects as the dust from aggregates. See the warning in the **AGGREGATES** section. Wear an approved dust mask.

Very fine sand, brick dust and **pumice powders** are other possibilities.

Metal filings can be added for colour or interest: **silver**, **copper**, **bronze**, **brass**, or **stainless steel**. Steel or iron will rust. Copper and its alloys give a bluish tint to concrete. Using too many metal filings can slow down the set time.

Pozzolans

Pozzolans are silicates that combine with the calcium hydroxide (the excess alkalinity or free lime) produced as the cement hydrates. This combination makes more cementitious material (calcium silicate hydrate). In this way pozzolans enhance workability (make the mix more plastic), increase strength and reduce permeability and drying shrinkage. They can slow down setting times, so these desirable characteristics usually take longer to develop. Some pozzolans are available in blended cements. These are the main pozzolans:

Silica fume

Silica fume is a silicate, a by-product of producing silicon metal or ferrosilicon alloys. It is a very fine dark grey powder. Particle sizes are 1/100th the size of Portland cement particles which makes it a highly reactive pozzolan, i.e. it uses more lime than fly ash. If used in the right proportions (about 8% by weight of the Portland cement) it can double the compressive strength of the concrete, reduce permeability and increase density. It can be difficult to handle as it is very fine, and makes the wet mix "sticky."

Metakaolin

Metakaolin, a kaolin-based clay product (also used at about 8% by weight of the Portland cement) can double the compressive strength of the concrete, lower permeability and increase density but also has advantages over silica fume: it is a plain white fine powder (unlike the dark silica fume), it makes the mix creamier, and is less sticky than silica fume. Particle size is smaller than cement but bigger than silica fume. At a 10% replacement of cement, it uses up 50% of the free lime. Metakaolin is a fairly new additive and may be difficult to find. Plain kaolin has not been heated and is not as reactive as metakaolin, and therefore not as effective.

Fly ash

Fly ash is a fine-grained glassy powder collected from the flue gases of coal-burning power plants. The largest use for this recycled material is in concrete. It ranges in colour from grey to black and closely resembles the volcanic ash that was used in the earliest concretes. Quality varies; consult the manufacturer for amounts used, usually up to 15% by weight of cement.

Blast furnace slag

Ground, granulated blast furnace slag (GGBFS) is another recycled pozzolan and can be used to replace part of the Portland cement (approximately 25% to 50%). Its benefits are similar to other pozzolans. But it also slows down the set time, which means it takes longer to get to final strength.

WARNING!

Pozzolans (silica fume, fly ash, etc.) are silicates. The inhaled dust over the long term can cause bronchitis, silicosis and lung cancer. Wear an approved respirator. Blast furnace slag dust can cause bronchitis, dermatitis and allergic dermatitis. Wear gloves and an approved respirator when using it.

Air Entrainers

Air entrainers are used in concrete that may be exposed to frost, or repeated freeze-thaw cycles. They create tiny air bubbles in the wet mix. MicroAir® is one brand. Very small quantities are used. (MicroAir: 1 oz. per 100 lbs. of cement. This is a 1:1,600 ratio. My calculations: 3.6 drops per lb.). Air entrainers need machine mixing; hand mixing is not enough to incorporate air bubbles in the mix. In some areas air-entraining cements may be available, such as type IA.

WARNING!

There are many types of air entrainers. Check MSDS and label for safety information.

Superplasticizers

In mixing concrete the object is to get all the dry particles in the mix to be "wetted" (forced into contact with the water.) The aim is to use as little water as possible, because water not chemically combined remains as porosity causing weakness and potential durability problems. Plasticizers (sometimes called superplasticizers or water-reducing admixtures) help to wet the surfaces of the cement, enabling the mix to use less water and still be workable. One brand is Pozzolith®, another Polyhede®. Very small amounts are used, and they're added to the water first. (Pozzolith: 2 to 5 oz. per 100 lbs. of cement. My calculations: 12 drops per lb.) Melamine (e.g. Melchem®) is another type of superplasticizer used to reduce water (about 1/2 tsp. per pound of cement).

WARNING!

Check MSDS and labels for safety information for various plasticizers.

Accelerators

Calcium chloride is used as an accelerator (about 1 tsp. per pound of cement) to shorten the set time – either because you are in a hurry or the temperature of the mix or the environment is cool. Dissolve it in water before using. Be aware that it also weakens the concrete and can corrode reinforcing. There are also accelerators that are not chloride based.

WARNING!

Calcium chloride can cause eye and skin irritation. Long-term exposure can lead to dermatitis.

Retardants

Retardants, or 'set retardants,' increase the set time of concrete. You might need them in extremely hot working conditions. Pozzolith® Retarder is one brand. (Master Builders, 3 to 7 oz. per 100 lbs. of cement.)

Latex or Acrylic

Acrylic and latex additives are polymer emulsions. Latex additive (also called latex polymer or latex admix) is a milky liquid sometimes used in concrete. It can be diluted with water. It also acts as a plasticizer (so you can use less water), increases strength and makes it more waterproof. It also increases adhesion – a useful property when adding a new concrete surface to an older concrete surface. Acrylic admixture (acrylic polymer) is an excellent alternative with similar properties.

WARNING!

Latex additive may be a skin irritant or an allergen for some people. Read the label. Acrylic admixture may cause skin irritation.

Mould Release Agents

Release agents are sprayed or painted onto moulds, so that the concrete releases more easily when set. There are many types, some specific to the mould material.

A mixture of castor oil and ethyl alcohol (denatured alcohol) in a ratio of 1 to 5, can be used as a release agent for **latex rubber** or **polysulfide rubber**. **Polyurethane** moulds need a special release, such as In and Out Release® made by Smooth-On, Inc. **Silicone rubber** moulds need no release agents for concrete, but a weak dishwashing solution helps reduce surface bubbles.

For **plaster** moulds or **metal** moulds (such as brass or steel) a paste floor wax or a mixture of melted stearine and kerosene (1 to 2) can be applied. On **metal** you can also use the castor oil and alcohol mixture.

Wood moulds should be sealed with shellac or a waterproof paint, then a paste wax or the stearine mix can be used as the release. Some artists report success with vegetable oil or mineral oil. Simple geometrically shaped mould parts can often be covered with plastic food wrap or thin plastic packing tape. **Fiberglass** molds need paste wax or the stearine mix.

This list is based on one developed by Douglas & Sturgess (artstuf.com).

5d

Reinforcement

Steel Reinforcing

Steel is the most commonly used reinforcing to give concrete tensile strength. It has almost the same expansion and contraction rate as concrete at differing temperatures.

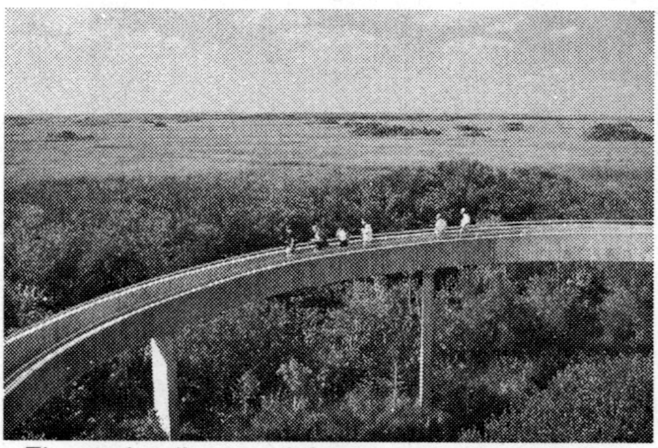

Elevated walkway: reinforced concrete. [Arttoday]

Steel rebar

In construction projects the steel used is known as rebar, and is sold as 1/2" to 1" (10, 15, 20 mm etc.) round rod. Smaller art projects can use much thinner steel rod or wire. The steel should be clean and have a rough surface for the cement to hold onto. Smooth rod for art projects should be roughened with a file or kinked to make a mechanical bond with the concrete. The rebar structure, or armature, is usually held together with steel or iron binding wire.

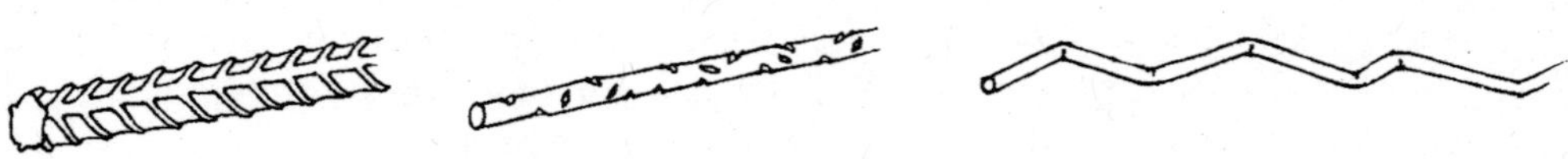

Left to right: rebar, thinner rod that has been filed, thinner rod that has been kinked.

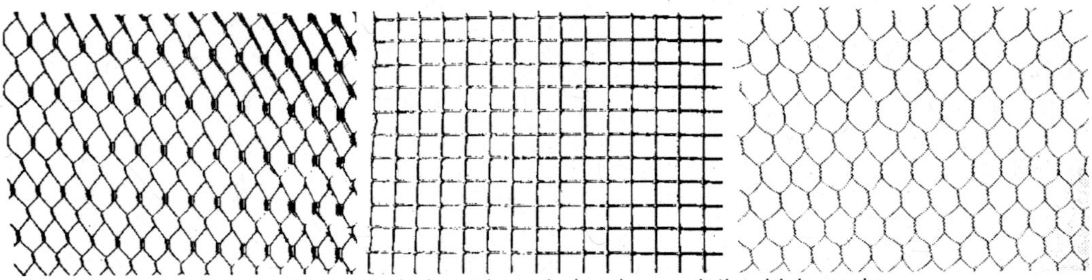

Left to right: expanded steel mesh, hardware cloth, chicken wire.

Steel mesh

There are also varieties of galvanized steel meshes or screens available that are suitable for smaller art projects. One type of expanded steel mesh (or expanded lath) is used as a stucco base and can be cut with metal snips. It is available in 30" X 96" sheets. There is also square grid galvanized mesh (or hardware cloth), 1/8", 1/4", 1/2" and so on, often used to make rabbit cages. Chicken wire, a lighter woven mesh used for fencing, is also a flexible and easy-to-use material. Meshes are held together at the joins with binding wire.

WARNING!

Cutting these steel meshes leaves very sharp edges. Wear gloves.

Fibres

Fibres are added to concrete for tensile strength, either for initial shrinkage-crack prevention, or to function in some way like steel reinforcing.

Plastic Fibres

Polypropylene fibre additives, such as FiberMesh® and Fiber-Ad®, are chopped strands of plastic about 1/2" (1 cm) long. They disperse into the wet concrete mix and help prevent shrinkage cracks as the concrete sets. Shrinkage cracks can cause problems later, such as larger cracks and water penetration. Ready-mix companies use it in a ratio of one pound per cubic metre of concrete. I estimate about 1 part fibres to 50 parts cement (not concrete) by volume. On very small-scale work the fibres also add a small amount of tensile strength. They are almost invisible in the final product though you may have to burn off the protruding fibres with a propane torch. **Nylon** fibres, such as Nycon®, are another alternative. They lie flatter and don't protrude.

Recent research in plastic fibres has led to **PVA** fibres or **polyvinyl alcohol** fibres. They appear to give much greater tensile strength than polypropylene or nylon because of the superior molecular bond between the fibre and the cement. They are less stretchy and stiffer than other plastic fibres and are now being recommended as a replacement for glass fibres (they are more alkali resistant) and in some cases, steel mesh.

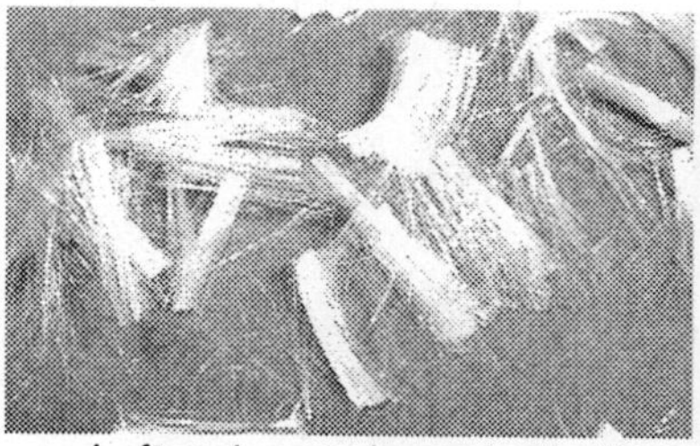
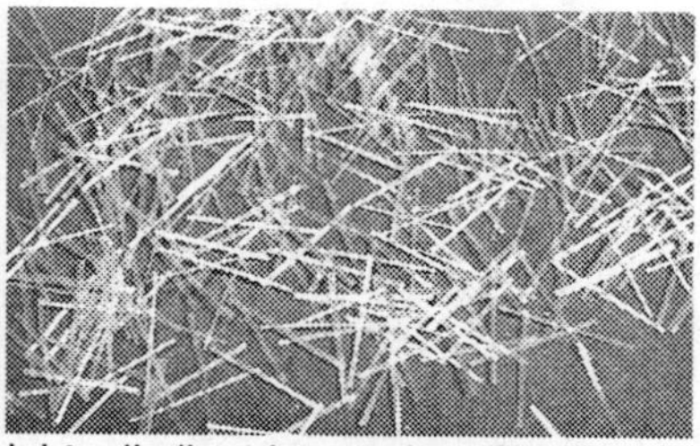

Left, polypropylene plastic fibres; right, alkali-resistant glass fibres.

Glass Fibres

Glass fibres added to the concrete mix must be alkali resistant (AR), or the alkalinity of the cement will eventually break them down. Their purpose is to add tensile strength to the concrete, replacing steel reinforcing in special circumstances. This makes them a useful addition for artwork in which the concrete might be cast in a space where steel reinforcing could not be used. The coated fibres, such as Cem-FIL®, are usually 12 mm long, and are added after the wet mix is ready. They are folded gently into the mix, not stirred vigorously, which could break the strands. The proportion of glass fibres is about 8% by weight of the cement (or 4% by weight of the total concrete mix). AR Glass fibres are also woven into a **fiberglass mesh** that can be used as a reinforcing material similar to the lighter steel meshes.

It is controversial whether to use AR glass fibres in concrete work used outdoors. Some long-term studies suggest that constant moisture over a long period of time, combined with the alkalinity of the cement, can break them down. (Source: Glass Fibre Reinforced Cement, A. J. Majumdar and V. Laws. See **BOOKS**.)

Other Fibres

Concrete has been reinforced with fibres like hair and straw since Roman times. Other fibres or strands have been used as reinforcing to increase tensile strength. Among them are **carbon** (sometimes used in concrete canoes) and **stainless steel**. Strands are typically 1" to 2" long and are crimped or bent at the ends. They are used at 0.5 to 2.0% by volume. Steel fibres increase impact resistance and tensile strength.

5e

Colour and Finish

Pigments and Dyes

Concrete can be coloured or dyed with a limited palette of colours. Because Portland cement is caustic, some pigments that work well in paint media are attacked by the cement, resulting in colour breakdown over time. To be certain, you must use pigments that have been proven to last when mixed with concrete.

The most typical concrete colours are earthy: brick or rust red, buff or yellow ochre, brown and charcoal (close to black). Green is less common and more expensive. Blue is even more expensive. (How often have you seen blue concrete?) Most colours seem to be derived from iron oxides or cobalt. A good colour chart can be seen on the Internet at: http://www.solomoncolors.com/concreteclrs.htm.

The colours typically come in a powder form, or mixed with water as a paste. Powder can be mixed in with the dry ingredients – although it's hard to see what depth of colour you are getting – or into the wet mix. Pastes are mixed into the mixed concrete.

Getting a deep rich colour can be a problem because most of the pigments have a maximum percentage (usually 10%) that can be used before they weaken the concrete. You can be left with a weak concrete with a good colour, or a strong concrete with a pastel shade. Using grey or white cement can change the quality and strength of the colours. The amount of pigment used is in the range of 1 to 6%. Often – for cost and strength reasons – the main body of concrete is mixed without the dye, and then a mix of cement and pigment is used as a top coat. This gives you the strength, with a deeper colour on the surface. The Arizona Oxides web site has a good overview at: http://www.arizonaoxides.com/concolor.htm.

Marbling or streaks of colour can be obtained by mixing two batches of concrete with different amounts of colour. They can then be partially stirred together and used. Or powdered pigments can be sprinkled on the wet mix and only partially stirred in.

WARNING!

Many artists' pigments are known to be toxic, particularly in the powdered state where they can be inhaled or ingested. Most concrete pigments come as a powder. Find out from the supplier what is in them. (Iron oxides are commonly used and have no serious ill effects. Cobalt compounds have a high toxicity rating.) So once again, use a dust mask and gloves when mixing concrete.

Acid Stains

Acid stains are painted on the surface of a completed concrete piece. The acid etches the surface, creates porosity and allows the stain to penetrate. Acid stains work well on old concrete. Because this is a technique used with finished concrete I have put more details in the **SURFACE FINISHING** chapter.

WARNING!

Acid stains are corrosive (hydrochloric acid). Read the MSDS and label carefully. Wear safety goggles, rubber gloves, and a protective apron.

Stains

You can also stain the surface of concrete using opaque water-based or acrylic stains. These stains are made to be used with outdoor wood, so there are no guarantees about long-term colourfastness when used on the surface of concrete exposed to the weather. Apply with a stiff brush and wipe off immediately with a slightly damp cloth. This leaves the colours embedded in the texture of the concrete.

Sealers

In most circumstances it is advisable to seal the concrete surface to reduce water penetration, to improve frost resistance and resist acid rain or staining. Sealing is usually the last step after colouring the surface.

Acrylic

Most of the readily available sealers are acrylic and water-based. Some may affect the surface finish, leaving it glossy. Also they may yellow slightly over time, and need renewing if used outside.

Siloxane

Silane or siloxane sealers combine chemically inside the concrete to make water-repellant UV-resistant silicone resin without changing the surface appearance of the concrete. Silanes penetrate more deeply than siloxanes. The solvents in some of these sealers demand a lot of ventilation (silanes: alcohol; siloxanes: petroleum-based). The surface of the concrete should be cured (at least a week) and dry.

WARNING!

Because these sealers are often solvent-based, you need a lot of ventilation when applying. Read the label carefully. Wear gloves, protect your eyes.

Potassium Silicate and Sodium Silicate

Sodium silicate, commonly known as water glass, is a traditional concrete sealer when mixed with water in a 1:4 ratio. It is painted on in several layers. V-seal® and other similar sealers (Seal-It® and Bio Vee®) are water-based potassium sealers for concrete and claim to be more effective over time than the siloxane or silane sealers. Both sodium silicate and potassium silicate sealers are clear and odourless and work by chemically combining with any free lime (calcium hydroxide) in the mix to form a gel. The gel hardens, making the concrete denser by decreasing porosity. If you are using pozzolans in your mix (see earlier discussion in Chapter 5c) most of the free lime may already be neutralized, so these sealers may not be needed. The concrete should be cured for a week, and damp (not wet or dry) when it is applied. If the concrete is of inferior quality or very porous this sealer may not work well and you might be better off with the older acrylic or vinyl sealers.

WARNING!

Although these sealers are water-based, clear and odourless, the label does caution about slight irritation to the skin and eyes, so wear rubber gloves and goggles. Ventilation is necessary.

Waxes

Some paste waxes work well as a final finish if you want a glossy surface. I look for ones that include carnauba for hardness.

WARNING!

Waxes can be a skin irritant and lead to dermatitis. Read the label.

5f

Premix

You can buy various concretes in the bag (e.g. Sakrete® or Quikrete®) where the dry ingredients are already premixed. If you only need small amounts this can be a good way to start, and you can improve the properties with some simple additives. Make sure the bag contains concrete, not plain cement. Premixes are usually sold in 60 or 80 lb. bags (25 or 40 kg).

Concrete Mix

This will be a mix of cement, sand and pea gravel (small pieces of stone). You may also be able to find these concrete mixes with added fibre for strength and fewer shrinkage cracks (see **REINFORCEMENT** chapter.)

Sand Mix

A mix of cement and sand, this is sometimes called a topping mix.

Mortar Mix

This is usually a mix of sand and grey Portland cement, with added lime. The lime makes the mix more slippery, but helps it retain water when applied to dry bricks and masonry. Lime also weakens the concrete, so keep this in mind if your work requires strength.

WARNING!

See the safety notes in the **CEMENT** and **AGGREGATES** chapters; the same cautions apply with premixes. Working with Portland cement, both wet and dry, and dust from aggregates, can have serious health effects. Wear gloves, an approved dust mask and eye protection.

Other Mixes

There are also fast-setting premixes and others made with lightweight aggregates. One manufacturer of pre-blended mixes lists five types, including a concrete mix, a sand mix, a fibre-reinforced mix, a fast-setting mix and a high-early strength mix – along with six mortar mixes including a coloured masonry mix.

6

Tools

Most of the tools for working in concrete you probably already have. Others you can buy easily at a hardware store or building supplier.

Mixing Tools

First you need containers to mix in. These can range from small plastic disposable tubs, to plastic buckets and dishpans, to 5-gallon food containers. For larger amounts you can use the finned mixers with the screw-on lid, shown here, or a wheelbarrow.

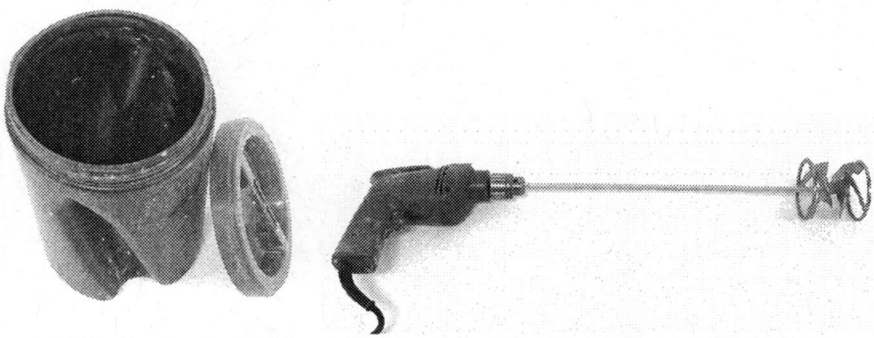

Mixing tools: plastic mixer with lid, power drill with paint mixer attachment.

For mixing batches for small-scale work you can use old knives, spatulas or hardwood sticks. I use strips of acrylic rod that I shape. For medium-scale mixing you can use small trowels, putty knives, or drywall tools. I often use a paint mixer, shown here in a power drill, used at low speed for mixes with air entrainers, and to disperse plastic fibres. I have also heard of using an old blender or electric kitchen mixer with a paddle attachment. For larger amounts use a hoe or a shovel with a wheelbarrow.

Application Tools

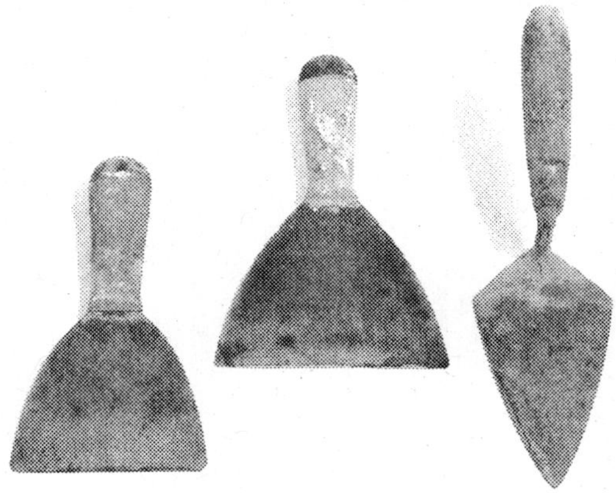

Trowels: two on left are for drywall and one on right for mortar

Some of the mixing tools work well for applying the concrete mix, such as trowels and drywall or putty knives. Strong rubber gloves work well for pushing the concrete into steel mesh and stopping it from coming out the other side. For applying small amounts of smooth finishing concrete mixes, thin vinyl or latex gloves work well. Don't use bare hands; read the **HAZARDS** chapter.

For levelling off the excess concrete at the top of a casting you need a straight edge of some sort. The edge of a trowel or a piece of wood works well for small castings.

If concrete is being placed in a mould, you need tampers to tamp the concrete in place. Pieces of wood with the ends cut flat, in a variety of sizes, work well. Pieces of steel (like cutoffs of rebar) can also be used to work the concrete into tight corners. Power palm sanders or muscle vibrators make handy vibrators applied to the outside of a mould when casting. Electric engravers, used for jewellery, work well as very small vibrators.

Carving Tools

For carving or texturing concrete that has not set, you can use almost any metal, wood or plastic modelling tools that would be used in clay. You can impress textures from any number of sources including screens or rubber matting.

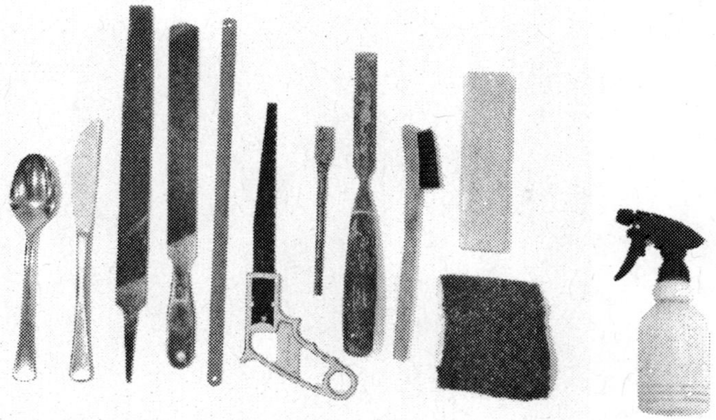

Carving tools, from left: spoon, knife, coarse files, hacksaw blade, small hacksaw, chisels, wire brush, coarse sandpaper with block. Spray bottle.

When the concrete sets, but can still be carved, you can use old knives, spoons, old rasps or wood files, chisels, saws, hacksaw blades, coarse grit sandpaper or garnet paper. Remember the sand in the concrete will dull your tools rapidly, so use old tools that don't matter, and clean them afterwards. Instead of files, I recommend the coarse sandpaper often used for refinishing hardwood floors (16 to 24 grit) as it can be thrown away when clogged with cement. A spray bottle for water is useful to have for misting work to keep it damp.

When the concrete is hard, after about two days, you can use stone carving tools. Grinding can be accomplished with various carborundum stones. For smoothing, you can use wet and dry sandpaper in grits going all the way to 400 and 600 for very smooth finishes. Using grinding stones and sandpapers wet makes them last longer and also produces less dust. Concrete drill bits, used with a power drill, can be used to make holes for mechanical connections.

Metal Tools

For sheet metal for moulds, you will need tin snips – the aviation type work well, or the traditional ones. You can also use these for cutting most of the finer meshes: the expanded mesh, galvanized meshes and chicken wire. Pliers, like needle nose or electrician's pliers, are used for bending the binding wire and tucking it in. Wire cutters are used for cutting the binding wire.

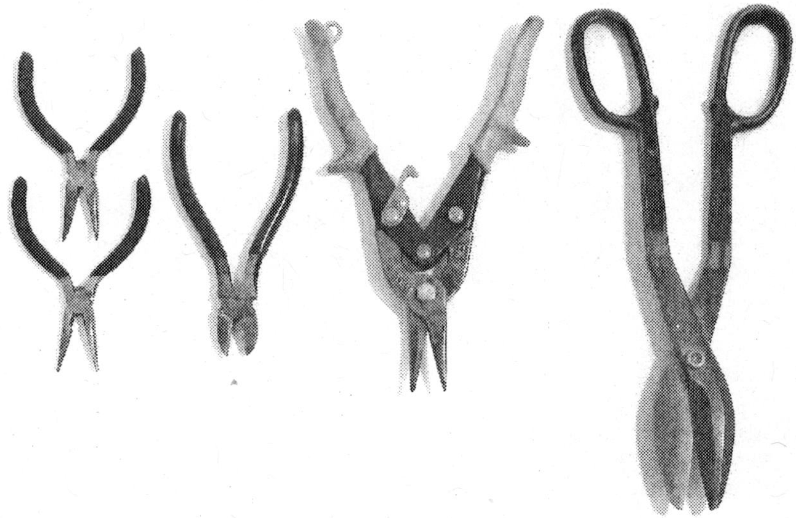

Metal tools, left to right: pliers, wire cutters, aviation snips, tin snips.

Other Supplies

Plastic sheet is good for covering the work while it sets to keep the water from evaporating. You can use drop sheets, plastic vapour barrier or food wrap plastic. Several varieties of tape are useful for holding moulds together, or sealing the plastic sheet. Make sure they are waterproof like duct tape or packing tape, clear or brown.

Safety Equipment

See **HAZARDS** chapter for what you need.

7

Mixes

The composition of the mix will be dictated by the size of the work, the method of forming, and its final use. Hard and fast rules don't always work. What follows are typical mixes for the applications suggested. You may have to experiment with several mixes before you get just the right one for a particular application.

As a general rule, the larger the final piece the larger the aggregate. For smaller pieces such as jewellery, you need a fine aggregate such as stone dust or finely screened sand. I use latex or acrylic admixtures as the liquid in small work because they toughen and waterproof the concrete.

For small sculptures regular sharp sand is the basic aggregate. In larger pieces, say sculpture in the 12" to 20" (30 to 50 cm) range, 1/4" crushed stone would give it strength and save cement. But if you wanted to carve the partially set concrete, the stone would be in the way, so just use sand as the aggregate.

In pieces that are to go outside in areas that have frost, you should use an air entrainer.

In all these mixes I recommend using metakaolin (or another pozzolan) for strength, and a plasticizer to cut back on the water in the mix. Although additives like these are not essential they do help to make a better, denser, stronger concrete than the old cement-sand-stone mixes.

With small mixes, when your measurements are accurate, you can add the cement to the water first, then mix in the aggregate. But usually you mix the dry ingredients first, then slowly add the liquid. Stir with a stick, spatula or a trowel – don't use soft wood that could split off into the mix. For small mixes for jewellery I use plastic containers; for larger mixes I use a bucket, sometimes mixing with a power drill fitted with a paint mixer at slow speed. Mixes with air entrainers need power mixing to develop the tiny air bubbles. For even larger mixes use a wheelbarrow and mix with a hoe or shovel. The new plastic barrels with internal fins that can be sealed and rolled on the floor work well.

So these are the basic mixes. They are ones I have used and developed over time and work well in terms of plasticity, compressive strength, density and finishing characteristics. You can change the quantities of any of these mixes as long as you keep the proportions approximately the same. Notice that I have not suggested liquid volumes because they are too variable (depending on the moisture content of the sand, etc.) but the general rule is to use as little as you can get away with, remembering that cement cannot combine with more than 40% of its own weight in water. Keep notes as you put together a particular mix; this will give you a future reference for the amount of water based on your own ingredients.

Volume Conversions

- 1 teaspoon (tsp.) = 5 ml
- 1 tablespoon (tbsp.) =15 ml
- 80 ml = 1/3 cup
- 175 ml = 2/3 cup
- 250 ml =1 cup
- 1 litre = 4 cups
- 1.5 litres = 6 cups
- 1 gram (g) = .035 ounces
- 1 ounce = 28.35 grams

Very Small Scale (jewellery)

- 1 tbsp. stone dust (screened limestone or marble)
- 1 tbsp. Portland cement (grey or white)
- 1/2 tsp. metakaolin
- pinch polypropylene or nylon or PVA fibres
- as small an amount of latex or acrylic admix (full strength) or water as possible (if you use PVA fibres, use water only)
- 1 tsp. metal filings optional (silver, bronze)

Mix the dry ingredients first, without the metal filings. Add the liquid slowly. The mix should be stiff, not runny. If you want to add metal filings do so after you've added the liquid and mixed it in. Then add the filings while mixing.

Mesh Sculpture (applied in layers)

This is a mix suitable for outdoor use, applied to an armature with steel mesh. Don't mix up more than you can apply in half an hour:

- 1.5 litres sand (sharp, brick or mortar sand)
- 1 litre Portland cement
- 175 ml stone dust (screened)
- 80 ml metakaolin
- 1/6 tsp. air entrainer (optional)
- 1/2 tsp. superplasticizer (optional)
- 1 to 2 tbsp. plastic fibres
- as small an amount of water as possible

Pull the fibres loosely apart. Again, mix dry ingredients first. Add the air entrainer and plasticizer to some of the water first. [Note: consult the manufacturer's recommendations for amounts of these additives. This mix is for MicroAir® and Pozzolith®.] Add this solution to the dry ingredients and mix, before adding more water. This way you know all the additives are in the mix. Add just enough more water to make a consistency that you can make a ball with, without it collapsing. If it's too dry the ball will crack apart; if too wet the ball will slump. The ideal would feel like modelling clay, although this is impossible to attain. Concrete is not nearly as plastic as clay. Now mix it some more to get the air entrainer to work and to disperse the fibres evenly. (You may find that adding the fibres to part of the liquid first helps to separate them. Then you would add this liquid mixture to the mixed dry ingredients.)

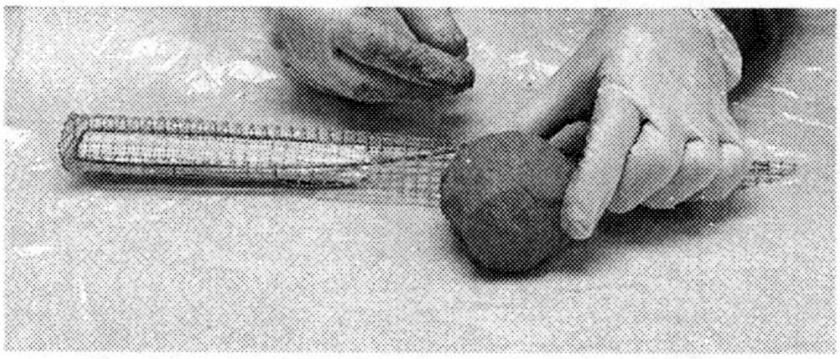

Correct consistency: the ball holds together.

Cast Sculpture (placed in a mould)

Basically this is the same mix as above, but you can add washed, crushed stone, the size of which depends on the size of the casting.

- 8 to 16 litres crushed stone
- 8 litres sand
- 4 litres Portland cement
- 0.5 litre metakaolin
- 80 ml fibres
- 1 tsp. air entrainer (optional)
- 4 tsp. superplasticizer (optional)
- as small an amount as possible of water

Pull the fibres apart loosely; mix dry ingredients together first. Add the air entrainer and plasticizer to some of the water. Add this solution to the dry ingredients and mix well, before adding more water. Add just enough water so that the mix is plastic and will flow properly into the spaces in the mould: the more details, the more fluid. The mix shouldn't be runny. It's better to push the concrete into place than have it too wet.

For comparison purposes, the Portland Cement Association in their 1995 Concrete Information Bulletin IS174, gives several mix proportions for what they describe as "small jobs" such as retaining walls and sidewalks. The mix for 3/8" aggregate, by volume, is: 1 part cement, 2.5 parts wet sand, 1.5 parts wet aggregate and 0.5 parts water.

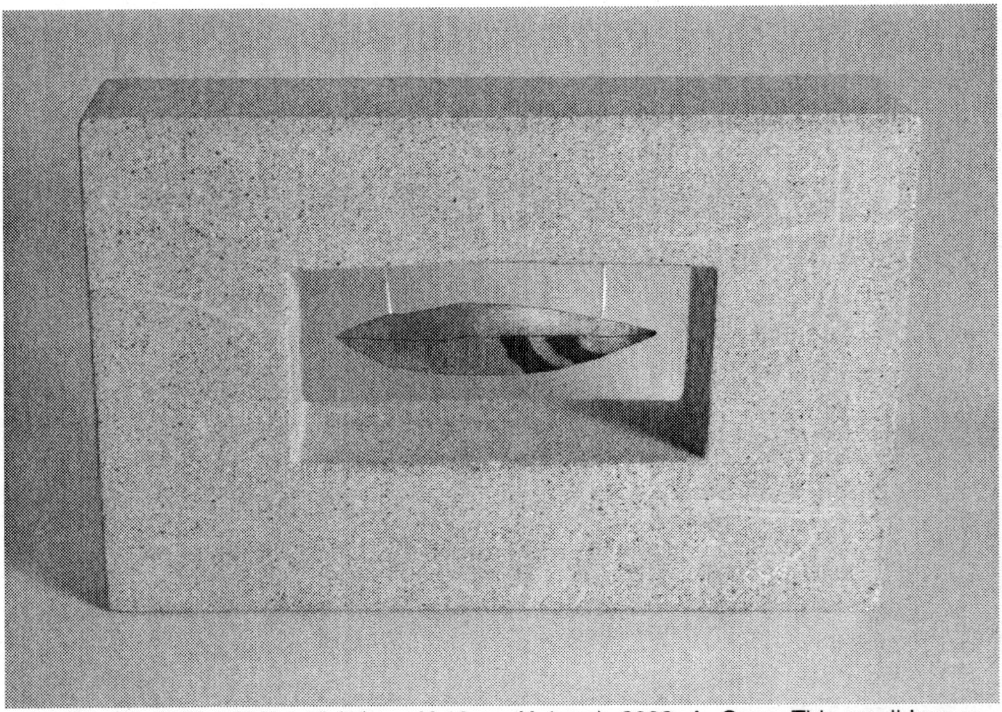

Feather #2: 7" X 4.5" X 1.5" (18 cm X 12 cm X 4 cm), 2002. A. Goss. This small bronze and gold leaf pin is held by silver wire embedded in a concrete frame. The concrete was cast in a wood and foam mould. The mix is the same proportion as the cast sculpture mix above, but without the stone.

Smooth Coatings

Use a smooth coating to rub into cracks or holes, or to overcoat as a final smooth layer. If you leave out the stone dust, then make sure this coating is not too deep or you may get shrinkage cracks.

- 300 ml Portland cement
- 150 ml stone dust (put through a screen – for very smooth you can leave this out)
- 40 ml metakaolin (if no stone dust, increase to 80 ml)
- as small an amount of latex or acrylic solution as possible

Hyper Tufa

Tufa is a natural porous stone. Hyper tufa is a cement mix made to imitate it. It's usually used for outdoor work that needs a rough, porous, stone-like effect. I have little experience with hyper tufa, but here is a summary of several mixes from other makers:

Hyper Tufa #1
- 1 part Portland cement
- 1 part sand
- 2 parts peat moss
- water or acrylic (latex) additive

Hyper Tufa #2
- 1 part cement
- 1 part peat
- 1 part perlite
- 1 part polypropylene fibres
- water or acrylic (latex) additive

Hyper Tufa #3
- 1 part cement
- 2 parts peat
- 1 part sand, vermiculite or perlite
- water or acrylic (latex) additive

http://whatcom.wsu.edu/ag/homehort/hypertufa.htm

Hyper Tufa #4
- 3 parts cement
- 4 parts peat
- 5 parts perlite
- water or acrylic (latex) additive

Also see Bill Loney's mix in the **GALLERY** chapter. If these variations are confusing, consider the variables and design your own mix. You must have cement, and in all of these mixes it is about one quarter of the total volume mix. The peat gives it an organic, unpredictable surface. Perlite makes it lighter in weight, and rough textured. Sand makes it strong.

Lightweight

Various aggregates are used to make concrete lightweight and easier to carve. If enough lightweight aggregate is used, the resulting concrete can be cut with a saw, drilled or nailed. The technology was developed in house construction for insulated roofing and walls. Lightweight mixes can be used to fill the space in a hollow reinforced structure. It can also be used to fill the inside space in an armature to prevent the next layers of regular concrete from falling through, and give it something to push against.

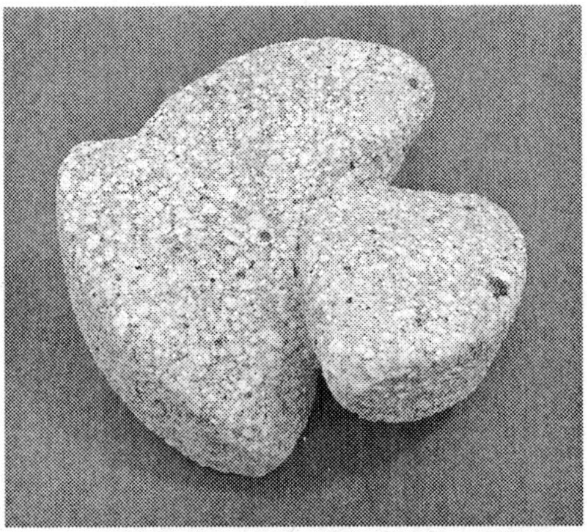

Partly carved perlite concrete test, about 6" across.

Perlite

Perlite concrete, used as an insulating roof material, can be made with a mix of:

- 1 part Portland cement
- 6 parts perlite
- water

This mix maintains a balance between insulation value and compressive strength for construction purposes. Other mixes range from 1:4 to 1:8. Sometimes sand is added for more strength along with an air-entraining agent.

Vermiculite

Vermiculite concrete is made for construction purposes with similar proportions to perlite:

- 1 part Portland cement
- 6 parts vermiculite
- water

One artist passed this mix along to me: equal amounts of Portland cement and vermiculite. And an alternate: 2:1:1 Portland cement: vermiculite: sand. These mixes may be too rich in cement.

Styrofoam Beads
Styrofoam beads can be used as a lightweight aggregate, but as mentioned before, there are difficulties to overcome. The surfaces of the beads are hard to wet down so the cement does not want to stick. A solution of polyvinyl propionate can be used to wet the surface. The light weight of the beads also tends to make them float in the mix, leading to poor dispersion. Shrinkage factors are high. One manufacturer makes coated beads to reduce these problems. See B-Lite™ in the **LINKS** chapter.

Lightweight Blocks
Lightweight concrete blocks are also made commercially, using proprietary chemical foaming agents. Air makes up to 80% of their volume.

Other Lightweight Aggregates
Pumice and glass spheres are also used to decrease the weight of concrete. See the **AGGREGATES** chapter.

Glass Fibres

Alkali-resistant glass fibres (ARGF) add tensile strength without the problems of steel reinforcing (rust, steel breaking through, thickness). There are however problems of long-term durability in a damp outdoor environment. ARGF concrete is particularly suitable for thin-walled pieces like bowls or planar forms.

This is a mix recommended as "general purpose" by Cem-Fil® (www.cem-fil.com), a manufacturer of alkali-resistant glass fibres. Note: these proportions are by weight, not volume:

- 50 kg cement
- 33 kg sand
- 16 kg water
- 0.5 kg superplasticizer
- 3.61 kg ARGF (12 mm length)

Dividing this by 50 we get more usable amounts:

- 1 kg cement
- 670 g sand
- 320 g water
- 10 g superplasticizer
- 72 g ARGF (12 mm length)

These are very cement-rich mixtures; for less shrinkage use equal weights of cement and sand. For small flat products use longer strand fibres (24 mm).

8

Setting (Hardening) and Curing

Several factors contribute to the time it takes for concrete to set. It is important to be able to control setting time because you want to know when you can pull the mould apart to do surface work, or add more layers.

The mix starts to set – gets too stiff to work with easily – after about half an hour when the cement starts to harden and bind the aggregate together. (If the cement paste starts to set after 5 to 10 minutes, this is probably a "false set." It can be safely remixed.) Normally you can break a mould apart, or turn an armature piece over, after about six hours (see variables below). This is what I call the "initial set." But test it first.

- If you scratch it with a nail and the nail easily penetrates, it's not set.
- If it scratches fairly easily with a nail, but has a crunch to it and feels hard, you have an initial set.
- If it scratches just on the surface and you can't dig in with the nail, it's gone beyond the initial set. The only way of shaping it at this point is with chisels or stone-grinding tools.

Experience is the only real teacher in deciding when it is safe to remove a mould.

Temperature

The warmer the mixture and the warmer the environment, the faster the set. This gives you two variables to control:

- By adjusting the temperature of the water added you can adjust the setting time, or more accurately the working time. In winter a ready-mix company will warm the water that's added to the concrete truck. In a smaller volume, the mix will adjust to the room temperature after an hour or two, but using water that is warm to the touch will shorten the set time.
- Be aware of the room temperature. I work in a cool basement most of the time (16°C, 60°F) so I usually wait about eight hours for the initial set. At room temperatures of 21°C (70°F) this time might be closer to five or six hours. At 30°C (86°F) the time will be even less. These times are rough averages for the mixes I use

with water at about room temperature. Recommended working temperature is 21°C (70°F) with recommended maximum of 33°C (90°F) and minimum of 10°C (50°F).

Water Quantity

I've emphasized repeatedly that you should use as little water as possible for increased strength. And the less water you use the faster the set time. A very stiff mixture with little water might set in four hours – a very wet mix might set in 12 hours. This is only for the initial set. After this first hardening there won't be too much difference in increased hardening (except for the fact that wet mixes generally produce weaker concrete.)

Additives

Some additives that increase the final strength of concrete (like metakaolin that can double the compressive strength of concrete) also appear to harden the concrete at a slightly faster rate. If you want to hand work a surface (wet carve it or file it) you have to take this into consideration. Other additives, like some of the pozzolans, may slow down the set and curing times.

More Additives

The concrete industry also uses accelerators (such as calcium chloride) to speed up the setting time or retarders to slow it down.

With all of these variables you have to test the concrete to see how much it has set. (See nail test on previous page.) You don't want to pull apart a mould and see the piece start to collapse. And at the other extreme you don't want to wait so long to pull a mould that you can't easily work the surface. Of course if you want to keep a cast surface as it is, with no extra refinishing, you can safely leave the piece in the mould, covered, for a week. At this point the concrete will be at about 50% of its maximum strength.

Curing

Although it is important not to use too much water in the mix, it is critical to keep the piece damp for at least a week, preferably longer. Concrete made with latex needs only a week's curing as the latex skins over, slowing evaporation of the water in the mix. Cement hardens by chemically combining with water, forming chemical compounds between the pieces of aggregate. As long as there is enough water present, the hydration reactions keep going, although at a diminishing rate, and the concrete keeps getting stronger.

If you let a piece air dry for the first week you will end up with a permanently weaker piece of concrete. Improper curing can yield a concrete that is only 40% of its optimum strength.

Keeping the work wrapped in clear plastic is an easy way of making sure the water in the mix does not evaporate. You can also lay a damp cloth over the piece. After a day check the work. If the surface is drying out, spray it with water and cover it again. Small pieces can be immersed in water after they have set, but make sure they have set. If you notice the cement dissolving into the water, it's too soon.

Once the concrete is curing it's too late to adjust the amount of water in the mix or the use of additives, but you *can* control the room temperature. A cold basement or studio will dictate a longer curing time than one week; a warm attic studio could mean a shorter curing time.

Keeping Notes

With variables such as air temperature, amount and temperature of water, pozzolans and other additives like accelerators and retarders, it is advisable to keep notes of your temperatures and mix proportions while you are measuring. This way you can predict the set time for future uses of the same mix. You can also duplicate the mix if you need to fill holes or add to the original piece.

9

Stresses and Strength

Stresses occur in materials either from outside forces, or internally from the weight or temperature differential of the material itself. A professional should engineer large sculptural works. The scope of this book is for small-scale work, but the same principles apply.

There are three types of force or stress on materials: compressive, tensile and shear. **Compressive** forces result in a decrease in size along the lines of the force. **Tensile** forces result in a lengthening along the direction of the force. And **shear** results in a sliding or slipping displacement of one part of a material relative to another part.

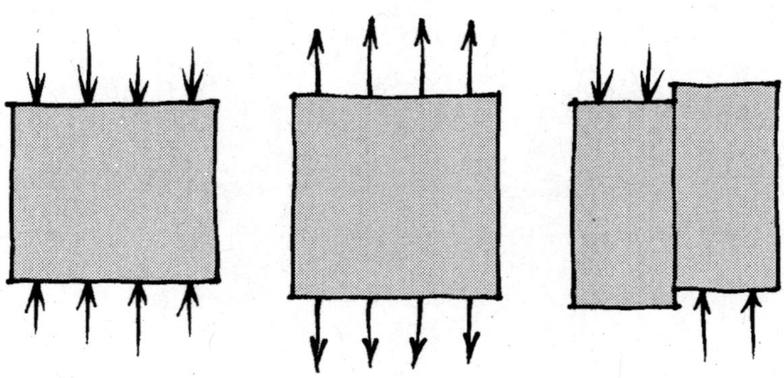

Three potential forces that can be exerted on concrete.

Concrete, like stone, is very strong in compressive strength (resistance to a squeezing force), but weak in tensile strength (resistance to being pulled apart). Concrete's low tensile strength means we have to add strength in tension. Steel is the common solution because it has high tensile strength, the same expansion and contraction rate as concrete with temperature change and is inexpensive and easy to obtain in a variety of forms.

The greatest tensile stresses in concrete occur near the outer surfaces as the following diagrams illustrate:

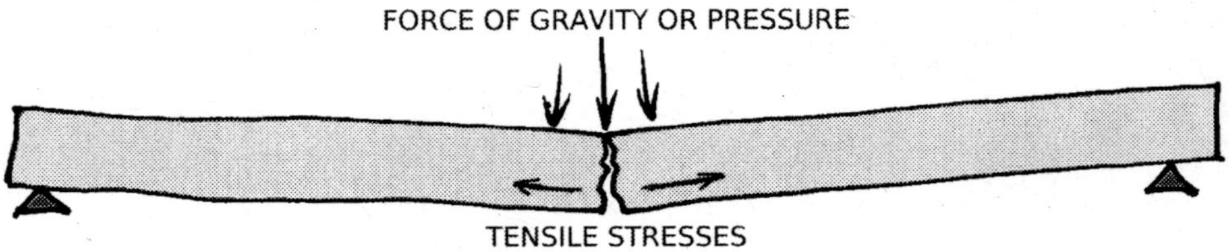

In this concrete beam supported at the ends, the weight of the concrete has produced tensile stress, causing it to pull apart and crack near the middle.

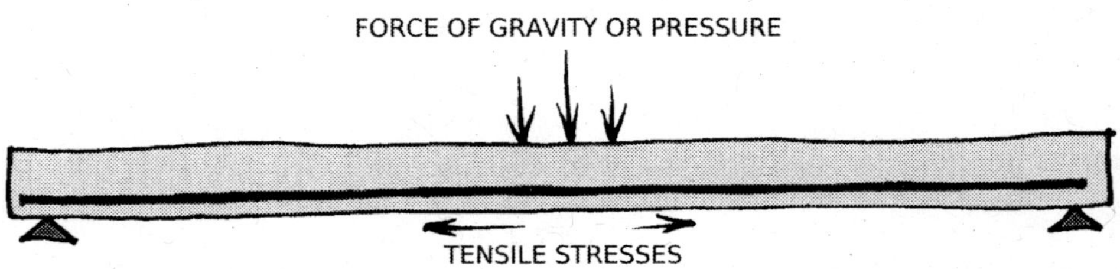

In this beam, steel reinforcement near the bottom surface adds tensile strength to counteract the stresses, stopping the concrete from being "pulled" apart.

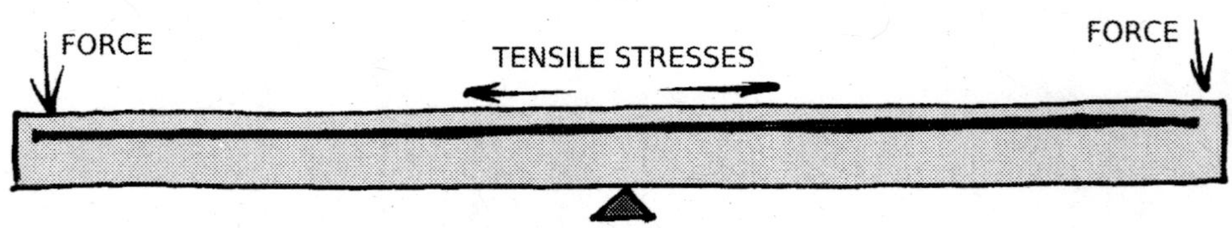

In this beam supported in the middle, steel reinforcement near the top surface adds tensile strength, stopping the concrete from being "pulled" apart.

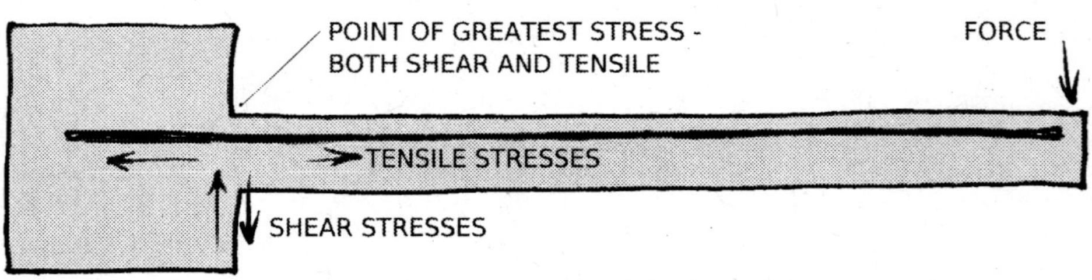

In this beam, held at one end, reinforcing should be placed near the top surface for both tensile and shear stresses.

Conclusions

In complex 3D forms you need reinforcing close to all surfaces. Make sure the steel is covered with enough concrete that it will not rust (at least 2 to 4 mm (1/8") in very small-scale work, at least 25 mm (1") in larger work). A heavy piece of rebar would do little in the middle of a beam, except counteract shear forces. Small closely placed pieces of steel wire near the surface create a more integrated ferro-cement, and are more efficient for counteracting tensile stresses. Consider where the highest stresses are when you are planning a piece: is it meant to hold a weight; will projections in the design cause stresses at certain points? Then place extra reinforcing at those points.

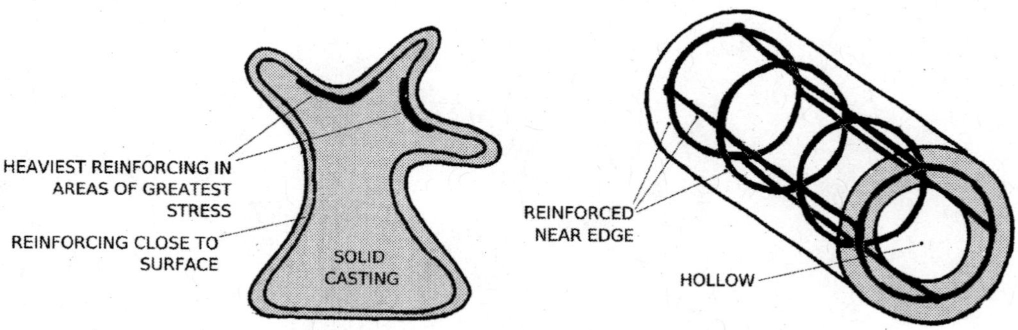

Left: solid construction. Right: Hollow construction

Temperature differences in a three-dimensional piece, particularly if placed outside, can cause stress as different layers of concrete expand and contract in sunshine, heat, rain or cold. For this problem a thin-wall reinforced hollow construction is advisable over a solid casting (see illustration above). In a solid casting the outer layers may be hot, for example, while the inside might still remain cool. The stresses caused by the relative expansion of the outside could cause cracking.

The following illustrations show a typical structure or armature used to create a small object – a knife-like form that is developed in a step-by-step sequence in the **MESH** chapter.

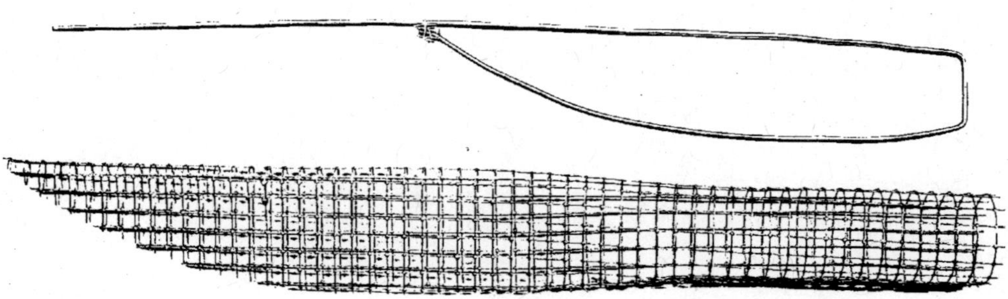

On the top, wire has been bent into the major reinforcing structure for the knife form. Below, the hardware cloth has been shaped for this form. The heavy wire will be placed inside the mesh form and tied in place with short lengths of wire. The handle part of the form will also contain a piece of shaped foam so that the concrete will be hollow when completed, with the reinforcing strengthening the skin of the piece. See the **MESH** chapter for a series of photos of the process from this point on.

10

Methods

a. Very Small Scale
b. Mesh: Step by Step
c. Casting: Simple Shapes
d. Surface Finishing

Methods of working in concrete vary with the scale of the work, surface qualities and form. The following sections explain how to work with concrete using several approaches: working at a very small scale, working with mesh and armatures for building up concrete, casting simple shapes using moulds, and ways of finishing the surface, including surface colour.

10a

Very Small Scale

Summary of methods

• constructing a structure out of metal or another material and using the concrete as an inlay
• making a mould from small pieces of metal, wood or plastic and casting concrete in that
• applying concrete to an armature
• making an original form in clay, plaster or modelling clay, and making a flexible mould from that using a variety of mould materials

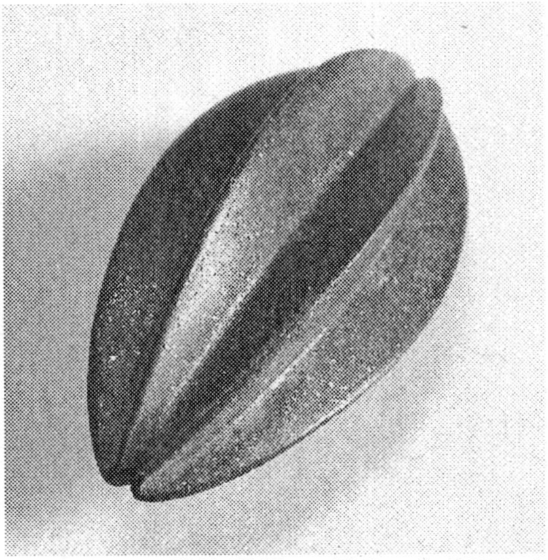

Pin: dyed concrete, 2" X 1" (5 cm X 2 cm), 2000. A. Goss.
This pin was cast in a brass mould, then wet sanded.

General Comments

Working with Portland cement for jewellery – like the pin shown here – I have developed mixtures that use very fine aggregates because the mix must be smooth enough to fill very small spaces. Even sand seems too coarse for some jewellery. There are no industry standards for making concrete jewellery, so these mixes are all from trial and error.

Why not just use Portland cement and water? It makes a very plastic, smooth material, but the shrinkage on setting is too great, creating shrinkage cracks or pulling away from the structure you are filling. For fine aggregates, the best is probably stone dust. You can get this as a waste product from limestone quarries or as marble dust. Unless the dust has already been screened, do so to get rid of larger particles. You can use easily available fly screen to do this, but I prefer even finer steel or brass mesh, or a fine mesh kitchen sieve. When you've screened the stone dust, use it with equal portions of Portland cement.

I also use metakaolin, an extremely fine white powder (see **MATERIALS** for other pozzolans), at about 8% of the amount of the Portland cement. Metakaolin can double the compressive strength of the concrete, making it denser and less permeable. Then I add polypropylene or PVA fibres to reduce the frequency of hairline cracks that tend to develop as the concrete sets. After setting, the protruding fibres can be burned off with a torch, being careful not to overheat the concrete surface.

For the liquid, substitute a latex or acrylic additive (made for concrete) for water. You can use it 50:50 with water, but using it full strength adds toughness to the mixture, and also makes it more waterproof, an important quality in jewellery.

General Purpose Mix

- 1 tbsp. (15 ml) limestone or marble stone dust (screened)
- 1 tbsp. (15 ml) grey or white Portland cement
- 1/2 tsp. (2 ml) metakaolin
- pinch polypropylene or PVA fibres
- as small an amount of latex or acrylic solution as possible or just water if using PVA fibres
- 1 tsp. (5 ml) metal filings (silver, bronze) (optional)

You can try leaving out some or all of the stone dust to make an extremely smooth, clay-like mix. But the design of the piece must be such that shrinkage doesn't matter, such as a small concrete piece held in place by mechanical means. If you are inlaying concrete into a metal cell (see diagram below) it might pull away from the walls of the structure as it shrinks. Or if it is to be cast on top of another rigid material, the shrinkage would show.

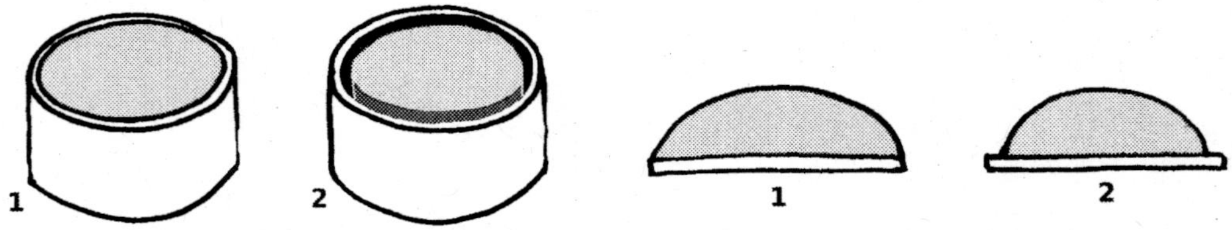

Left: aggregate-free mix in a tube of metal. When it sets it could shrink and pull away from the inside edge. Right: aggregate-free mix on a metal plate, side view. When it sets the shrinkage would be noticeable along the edge.

Mix the dry ingredients first. Add the fibres, then the liquid, as small an amount as possible. Mix this well – it should be thoroughly wetted, but not liquid. The consistency should be such that it does not slump. Place it in a walled form, or a mould, or around an armature depending on the method chosen.

Processes

Walled Forms (Inlay)

Although this is not inlay in the usual sense (where pieces are cut and fit into cells) it does give the appearance of inlay when finished. A walled form is constructed in metal, the walls textured on the inside for a mechanical bond. You may want to add reinforcement at this stage if the inlay covers a large area. Next, the mix described above is pushed in hard enough to remove air bubbles that might be trapped. Overfill the cement mix in the form. Cover with plastic wrap for about 24 hours, making sure the surface does not get dry. Remove the plastic wrap and wet sand the surface flat. Small air holes can now be filled with a mix of the same proportions and covered for another day, then wet sanded again. If any tiny holes are still visible they can be filled with a paste of cement and liquid, covered again, and wet sanded again.

After a week's curing (covered or submerged in water; not allowed to dry out) the concrete can be wet sanded with very fine wet and dry sandpaper (600 grit) and sealed. This curing period is important. Remember that cement hardens and cures by chemically combining with water. If it is allowed to dry out while curing it will be weaker.

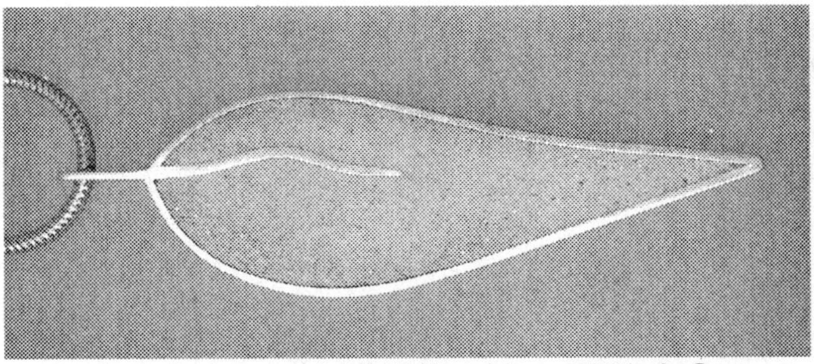

Leaf pendant: concrete, 2.5" (6 cm) long, 2001. A. Goss.

The leaf pendant shown here was made using concrete as an inlay material. The silver structure was built using 1 mm X 5 mm silver. The inside of the walls had grooves ground into them for more of a mechanical bond. Holes were drilled in the leaf's vein to accept steel reinforcing wire that was buried in the concrete mix. The concrete mix was then packed into the "cell."

Moulds

Simple moulds can be made from sheet metal cut with tin snips or plastic food containers cut with scissors. At this scale the moulds do not need to be strong. The pieces of the mould can be held together using tape: transparent tape or packing tape works well. Connections for moulded work must either fit the shape (as in a grooved form), or imbedded in the casting. The cement mix described above can be packed into the mould, being careful not to trap air. Overfill the cement mix in the mould. Cover with plastic wrap for about 24 hours, making sure the surface does not get dry. Then remove the plastic and carefully take the mould apart. Wet sand the surfaces. Small air holes can now be filled with a mix of the same proportions and wrapped up for another day, then wet sanded again. If any tiny holes are still visible they can be filled with a paste of cement and liquid, covered again, and wet sanded again. After a week's curing (covered or submerged in water; not allowed to dry out) the concrete can be wet sanded with very fine wet and dry sandpaper (600 grit) and sealed.

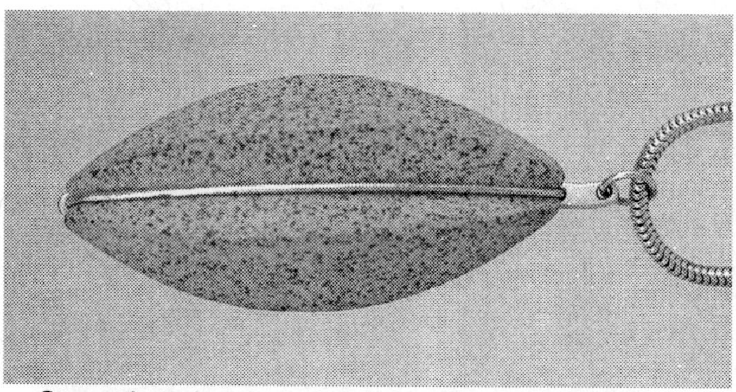

Grooved pendant: concrete, bronze filings. 2001. A. Goss.

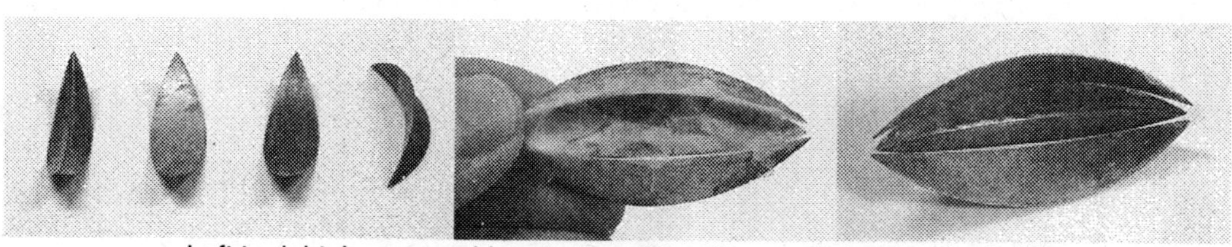

Left to right: brass mould parts, taped together, and showing the inside.

The grooved pendant shown above was cast in a small three-part break-apart brass mould. The finished concrete was then "strapped" to a silver back using silver wire.

Shrinkage isn't critical in this piece where the concrete is independent of the metal. The concrete was made using white Portland cement with bronze filings and stone dust as the aggregate.

Armatures

Steel wire and mesh armatures to make larger objects are covered extensively in the next chapter. For making jewellery and very small-scale objects, the same principles apply. A basic shape can be formed using finer meshes such as a 1/8" grid. Mesh can be held in position and joined using thin binding wire. With smaller dimensions it is not as important to make hollow forms, because temperature differentials are not a problem.

This concrete bracelet below has a silver band as the main structure. Silver pegs were drilled and then soldered around the circle. Stainless steel wire for reinforcing was threaded through the holes and wrapped around the pegs. A mixture of Portland cement, stone dust, silver filings, polypropylene fibres and latex solution was pushed into and around the armature. The concrete was supported on the inside surface of a bowl (a handy concave surface) while it set. Filing and wet sanding the concrete after two days completed the rough shape of the piece; small holes were filled; final wet sanding and sealing with a siloxane sealer took place about a week later.

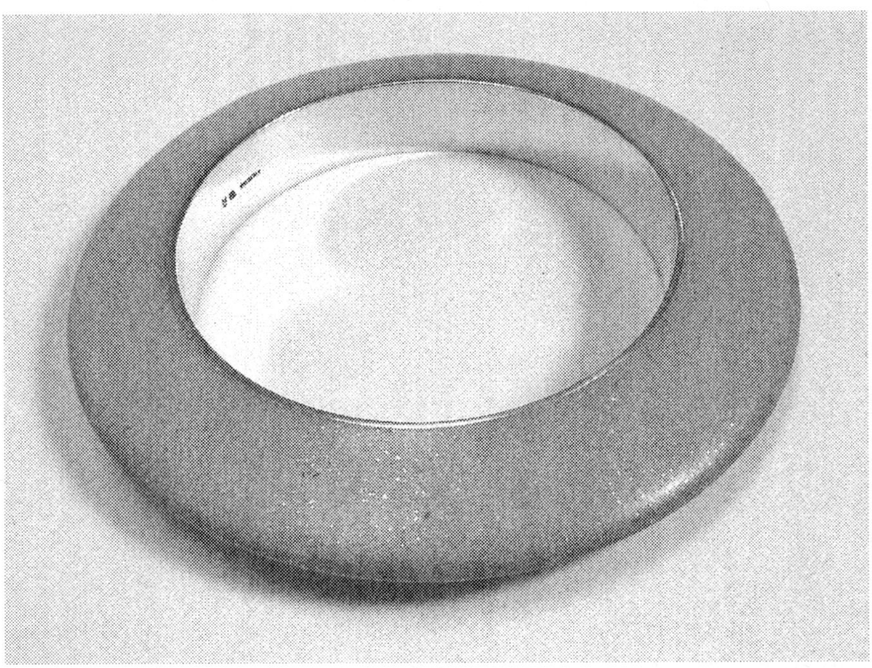

Bracelet: concrete and sterling silver. 1999. A. Goss

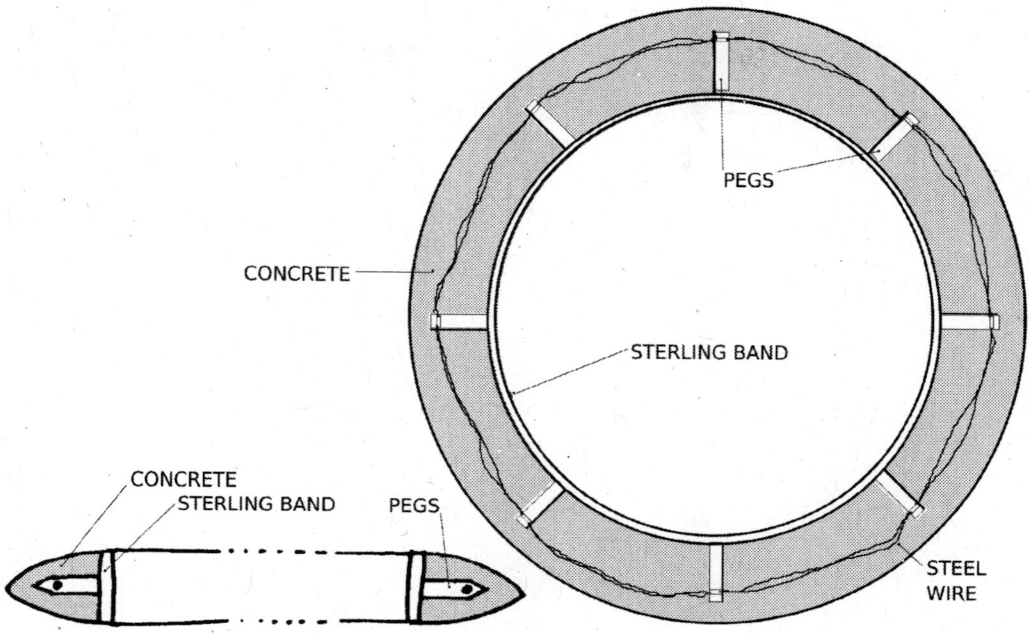

Left, cross section diagram of bracelet showing pegs and holes for wire.
Right, cross-section showing silver band with pegs and wire for reinforcing.

Sealing

When the concrete has set and cured, sealing can be done with a variety of sealers. I use water-based potassium sealers like V-seal®, applying it with a brush to the concrete (damp, not wet) and then wrapping the piece with plastic wrap for 24 hours. This allows the sealer to penetrate without evaporating. These potassium-based sealers chemically combine with any free lime in the mix to make a gel, which then hardens to produce a less porous and denser concrete. If using siloxane or acrylic sealers the surface should be dry. After drying, the piece can also be waxed, using a hard, carnauba-based wax. More details in the **SURFACE FINISHING** chapter.

Caution

As you are probably aware, concrete as a jewellery material is not common. This is part of the reason I find it an interesting and challenging material. But there are no guarantees about its longevity when in contact with oils and salts from the skin – although I have pieces that have been worn for several years and look fine. One would expect durability and hardness to be about the same as limestone. There may also be some health concerns about skin being in constant contact with concrete. I design jewellery to avoid this, for example placing metal between the skin and the concrete (see the bracelet example above). It is also important that the concrete be dense and impermeable. The use of pozzolans in the mixture is recommended as they both increase density and neutralize a lot of the free lime (causticity) in the cement. Sealing and waxing also make the concrete more impermeable to moisture.

10b

Mesh: Step by Step

This section shows how to apply layers of concrete onto the *outside* of a pre-made form. Various materials can be used to rough out the three-dimensional form:

- Carving a block of expanded polystyrene, foamboard or Styrofoam®
- Cutting and taping together pieces of cardboard
- Cutting and taping together cut pieces of corrugated plastic (Coroplast®)

Steel mesh is then cut to wrap or enclose the form, tied together with wire. Concrete is applied in successive layers, gradually refining the shape. This was the technique used in the concrete portion of the sculpture below.

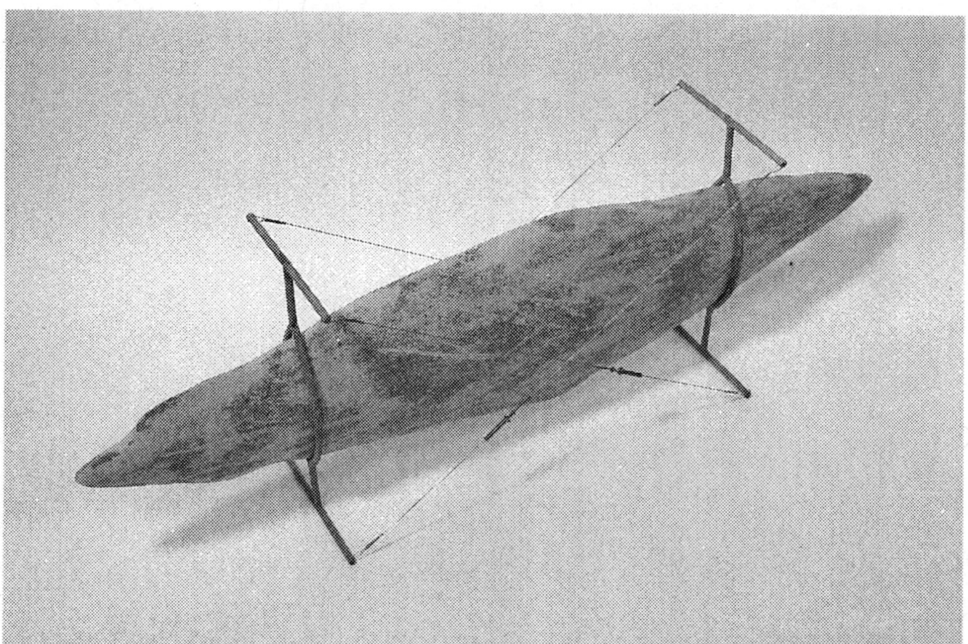

Cloud-Fish: concrete, patinated bronze, steel wire.
10" X 31" X 6" (25 cm X 79 cm X 15 cm), 1999. A. Goss.

Process #1

First, parts are cut out of plastic board to make a hollow three-dimensional form. Old election campaign signs work well. Strong cardboard can also be used. This method is particularly suited to making forms that have strong planes.

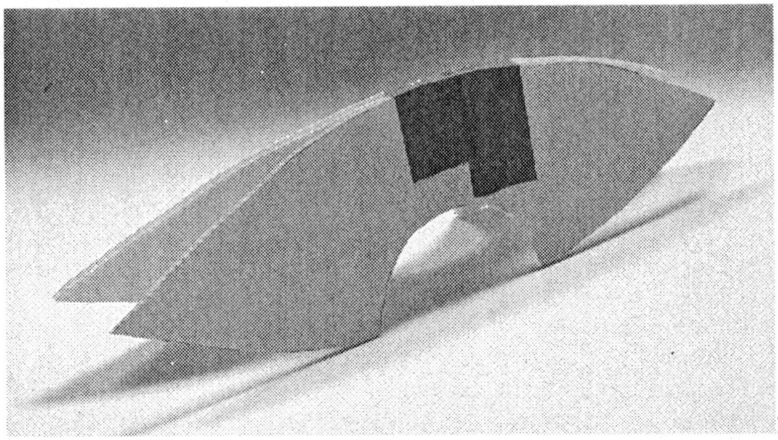

1

The parts are taped together into a three-dimensional form using packing tape, masking tape or duct tape. (1) The corrugated plastic is easy to bend into curves and hold with more tape. If you use cardboard you must completely cover it in waterproof tape. Remember that the structure has to hold together, while wet, for about 24 hours until the concrete sets.

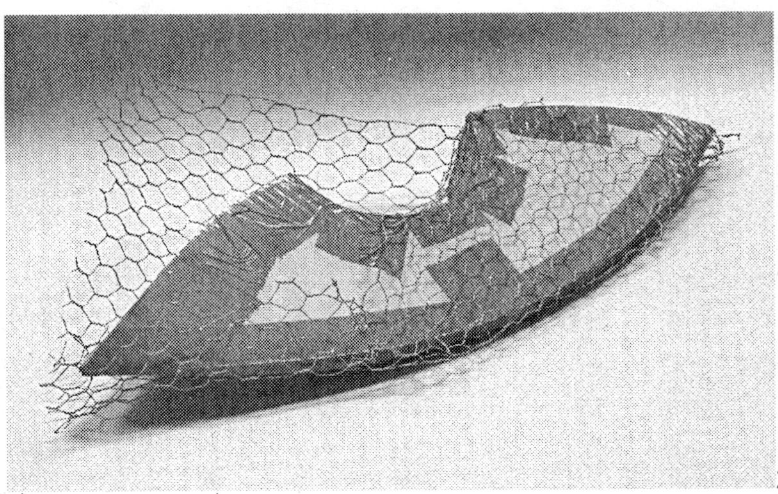

2

Steel mesh is wrapped around the form, which is now a hollow three-dimensional shape. (2) You can use chicken wire, expanded mesh or hardware cloth. The joins are tied together with binding wire. Loose ends are tucked in using pliers. With the thinner mesh like this chicken wire, you may need two or three layers for strength. (3)

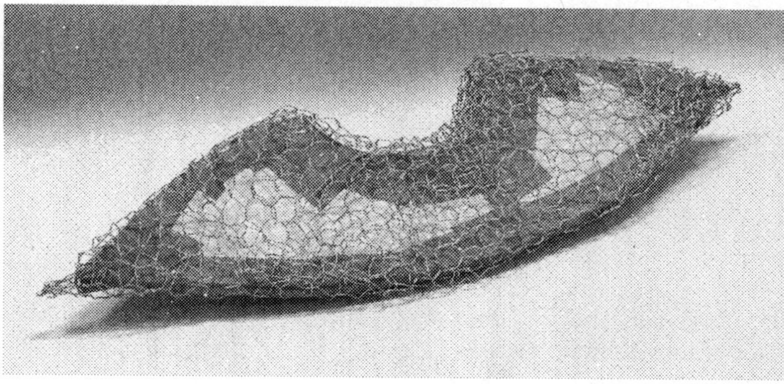

3

Mix the concrete – see Mesh Sculpture mix in the **MIX** chapter. Apply it by hand. Wear rubber gloves (please read the **HAZARDS** and **MATERIALS** chapters if you haven't already done so!) and push it through and around the mesh. You want a mix that is clay-like; if you make a ball with it, the ball should hold together, not slump too much. If it's too wet or too dry the mix will not stick to the wire. Remember that if the mix is too wet it will be weak. The wire must be completely surrounded and in contact with the concrete. Overlap each section of wet concrete you apply, so the join-line areas are well mashed together.

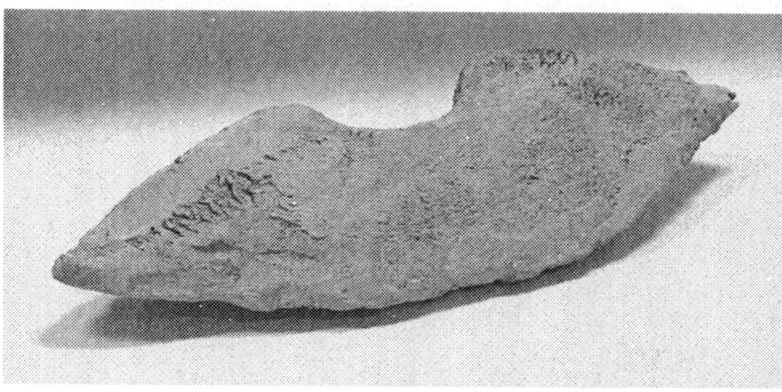

4

The concrete is applied to one side of the form and should be about .5 cm to 1 cm (1/2") thick. (4) It is difficult to get the concrete to hold upside down, so after completing the top, side and most of the vertical parts, leave the piece for 12 to 24 hours. Cover it in plastic to prevent water loss.

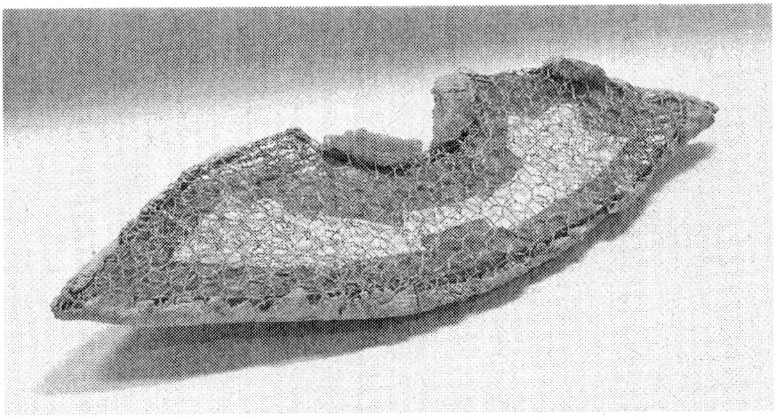

5

After the concrete has set on that first side you may want to rough sand or file it. If it looks dry, spray it with a mist of water. Flip the piece over – it looks like this. (5) Now mix up more concrete and apply it to this side. Cover and let it set for another 12 to 24 hours.

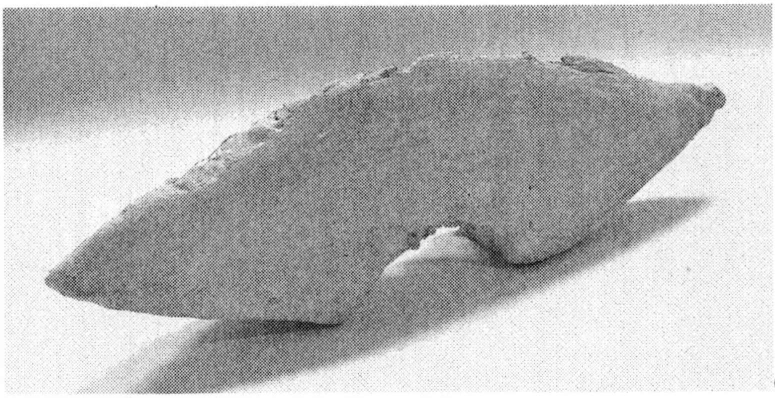

6

The first layer has been applied to both sides. (6) Rough sand (16 to 24 grit) or file it with a rasp to obtain the rough shape you want. At this point it should be hard enough not to break and soft enough to sand. (Wear gloves and a mask.) Don't let the concrete dry out: keep it damp with a mist of water from a spray bottle.

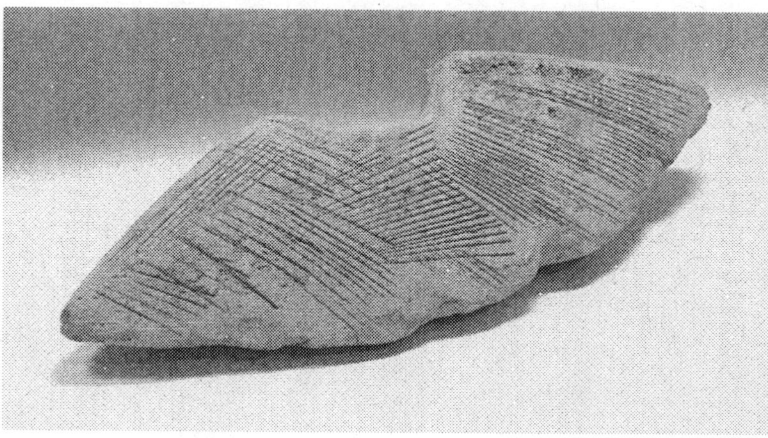

7

After sanding, wash off the concrete to get rid of the dust, remove excess water and apply a second layer of concrete (.5 to 1 cm) to the top surfaces. (If you have waited more than a couple of days, you can increase adhesion of the second layer by painting on a slurry of Portland cement and latex solution before adding the second layer.) This second layer can be textured before or after the set, to provide "tooth" for the third layer, and visual interest. Let it set for 12 to 24 hours, covered. The second layer means the concrete is now 1 to 2 cm thick on the one side. After it has set, sand it, flip the piece, and apply the second layer to the reverse side. Again, the surface can be textured. The form should now be very close to the final shape you want.

Mix up a third and final smooth concrete mix (see Smooth Coating in the **MIXES** section) if you want a dense or polished surface. This final mix can be a different colour. Apply this mix to one side with a trowel, spatula or smooth rubber gloves. You may be able to do all sides at once with this thin layer. Let it set. Except for filling minor blemishes you are now finished applying concrete. See also the **SURFACE FINISHING** chapter for details about adding material.

Wet sand all surfaces with a wet/dry sand paper, emery paper, or corundum paper (120 through to 400 grit) if you want a smooth surface. You can sand through parts of the last layer to reveal the base colour underneath. If fibres show above the surface at this stage, you can burn them off with a small propane torch, being careful not to get the concrete too hot. The piece should then be covered and kept damp for a week or longer, then sanded again (400 or 600 grit) to remove stains, and then dried and sealed.

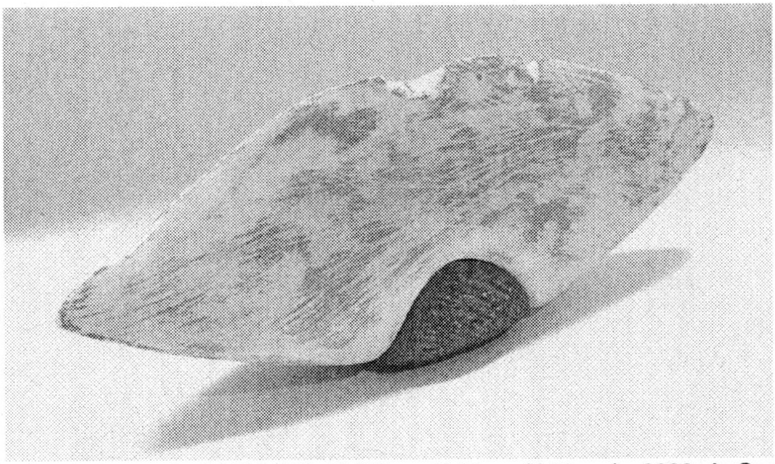

Sculpture: concrete, 7" X 21" X 4" (17 cm X 54 cm X 10 cm) , 2000. A. Goss

In the completed piece shown here I used a white final smooth coat over a grey body and sanded through the white to reveal tool marks. Then I cast a small base for the sculpture – the dark piece – and later joined the two pieces together with silicone adhesive.

Process #2

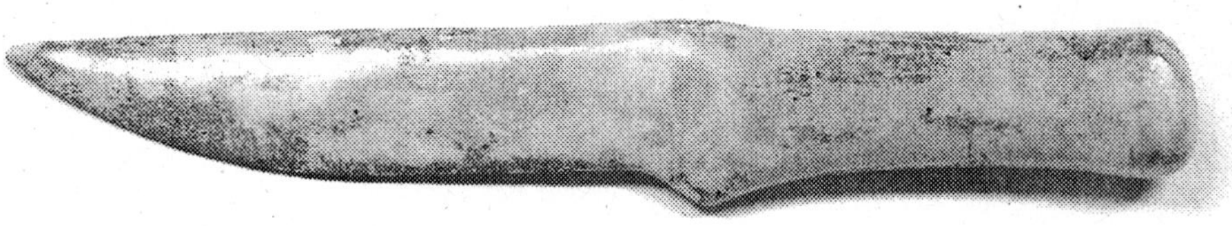

Concrete Knife: 14" (35 cm), 2002. A. Goss

The following series of photographs illustrates the steps in applying a smooth mix to a smaller armature. This knife shape has a flat blade area and a round handle. I made it as a technical challenge, but I also like the aesthetic of making such an object in an unexpected material. The handle has foam in the middle to create a hollow core, with the reinforcing mesh running close to the surface. A stronger piece of steel rod traces the edge of the armature and gives it rigidity. (See **STRESSES AND STRENGTH** chapter for an illustration.) The mesh is wired to itself and to the rod with pieces of wire. Sharp ends are bent under.

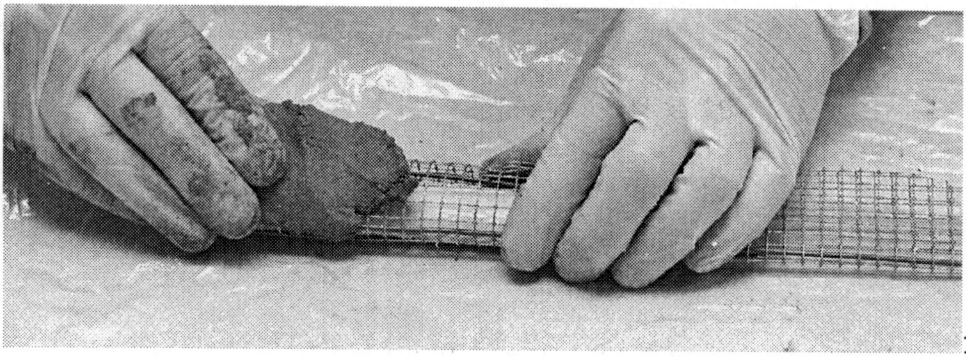

1

Apply a smooth concrete to the small mesh form by pushing it well into the wire to completely cover and surround it (1). The foam core stops the mix from being pushed to the other side. On the blade part you must use your own fingers to stop the mix from being pushed through.

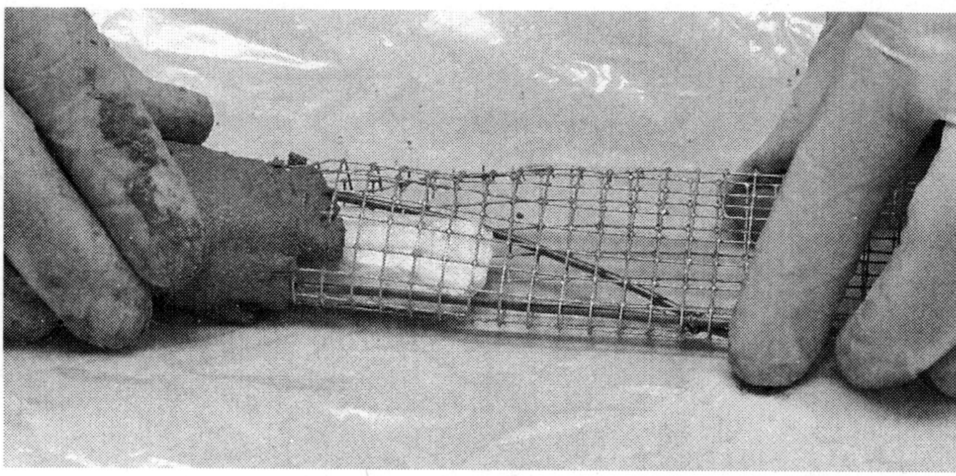

2

More detail (2) showing the mesh, the steel rod, foam, and the concrete pushed into place. Try and cover all the wire in this first step.

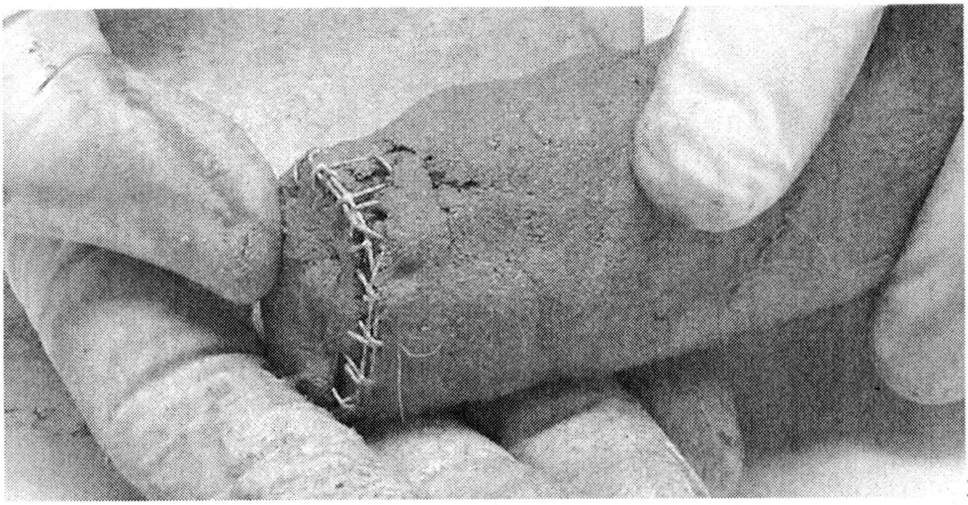

3

Close up (3) showing how the concrete is pushed through and under the wire mesh. Try and get rid of all air spaces so all the steel is contacted by concrete on all sides.

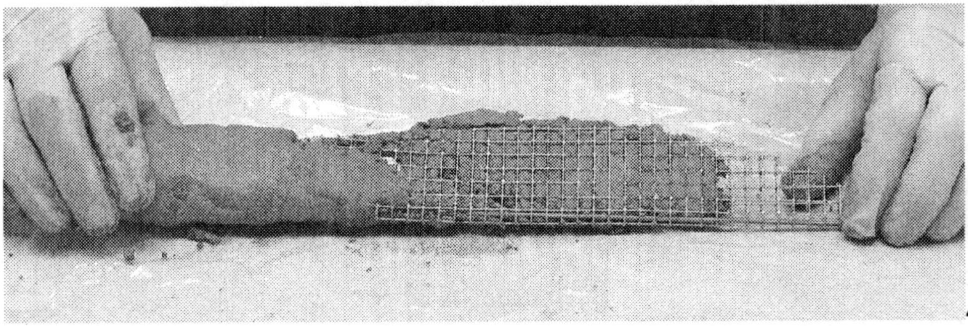

4

Finish applying the concrete to the "blade" part of the form (4). Notice how the mesh is almost buried on the one side.

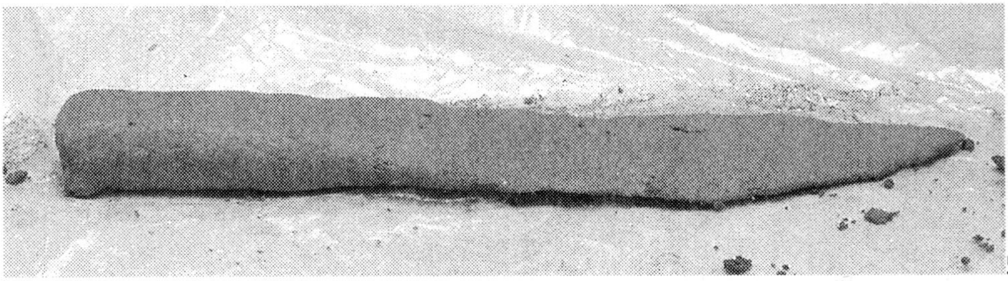

5

Completely coat the armature with concrete on all sides (5). No steel shows through. If your mix is the right consistency you can do all sides in this first procedure. If it's too dry it won't contact the mesh properly; if too wet it will fall off.

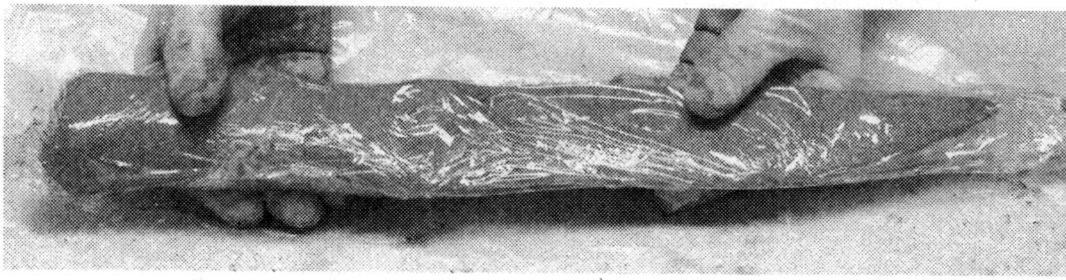
6

After more material is added to the handle, the form is wrapped in plastic wrap (6) to let it set for 12 to 24 hours. You want to remove the plastic when the concrete is:

1. hard enough to be resistant to breaking and
2. still soft enough to grind easily.

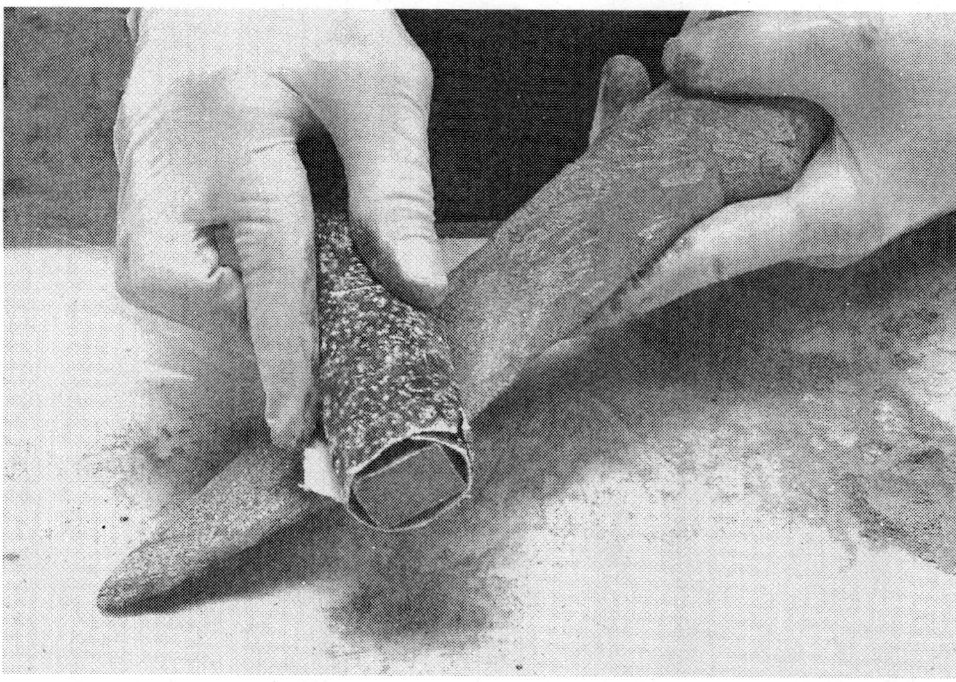
7

After this initial set, use a 16 to 24 grit heavy sandpaper wrapped around a piece of wood to shape and refine the form. (7) You are trying to define the form clearly without worrying about surface finish

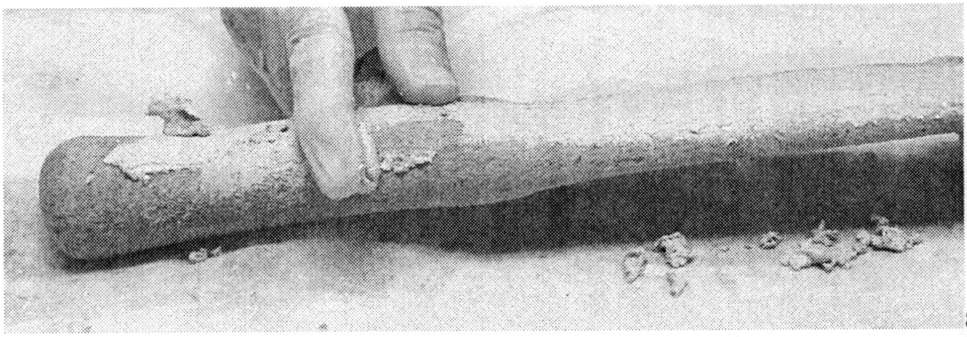

8

After dampening the surface (not wet – you are just trying to stop the concrete from pulling moisture from the new mix) you can add a smoother mix by rubbing it into the surface with rubber gloves. (8) Apply enough pressure to force the mix into any holes or crevices. Larger pieces might need a small trowel or a flexible putty knife to do this. When completely covered with this smooth mix, wrap again in plastic for 12 to 24 hours making sure after the first few hours that it is damp enough.

After wet sanding with 120 to 180 grit any protruding fibres can be burned off with a small propane torch, being careful not to overheat the surface. Then the piece is wrapped and kept damp for several more days, wet sanded with 400 grit, dried, sealed and waxed.

Connections

In all concrete work, connections to other materials or other concrete pieces must be planned in advance. There are several possibilities:

- Shape the work itself so that a metal or other structural piece will fit in a groove or some other shape to provide a mechanical connection. See *Cloud-Fish* sculpture at the beginning of this chapter or *Grooved Pendant* on page 61.
- Cast or imbed metal fittings in place. These could be bolts, threaded tubes, eyehooks, or solid metal that could later be drilled or tapped. Brass, bronze, stainless steel or galvanized steel are all suitable materials.
- Cast or imbed small blocks or cylinders of plastic foam in place. Later these can be cut out and the holes used for mechanical connections held in place with glue or cement paste.
- Concrete objects sitting on furniture may need protective bases such as cork or rubber. Use contact cement.
- For joining concrete parts together, use imbedded metal connections described above or adhesives such as contact cement, silicone or epoxy glue.
- After concrete has set, a concrete bit can be used in a power drill to drill holes. A connecting device such as a bolt could then be epoxy-glued into the hole. (Casting the bolt in place is a stronger alternative than this.)

10c

Casting: Simple Shapes

This is a description of simple casting. These processes could be used for casting shapes that are to be carved, or for final simple shapes. Casting of detailed, reproducible forms in plaster or flexible moulds is a highly specialized area and will not be covered in this book. (See **BOOKS** section for further references.)

The basic principle is to create a negative space that is filled with concrete. Any number of materials can be used that have the following characteristics:

- strong enough to resist the weight of wet concrete
- waterproof or able to be waterproofed
- easily assembled and disassembled when the concrete has set, or flexible

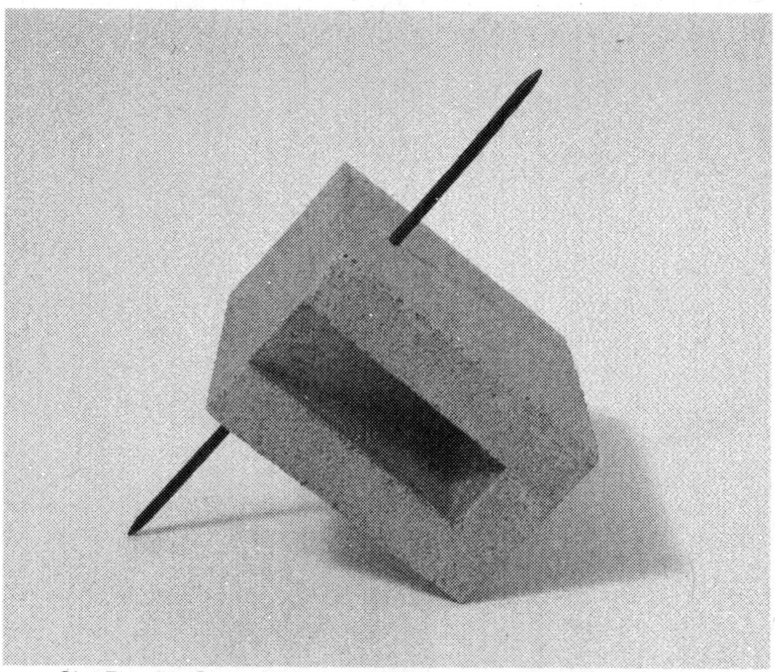

Slot Box #2. Concrete, about 6" (16 cm) across,1996. A. Goss.

A variety of mixes are used for different purposes. A cement and sand mix (1:1 or 1:2) can be used for the basic cast shape, followed by a cement mix with no aggregate or a very fine admixture (such as stone dust or metakaolin) for patching and finishing. [See the **MIXES** chapter for more details.]

Foamboard

For casting geometric forms like these, I use expanded polystyrene insulation, also known as Styrofoam®, or foamboard. It can be cut quickly with a sharp utility knife, or accurately with a scroll saw with a fine blade. This method is particularly suited to simple geometric forms. In the diagram below, five layers of 1" (2.2 cm) polystyrene are sandwiched together using double-sided tape and duct tape on the outside to hold it all together. Adhesive made for foam insulation can also be used. A forged copper rod was cast in place. When you are embedding metal or other materials in concrete always make sure there is a mechanical bond; in this case I filed grooves in the copper.

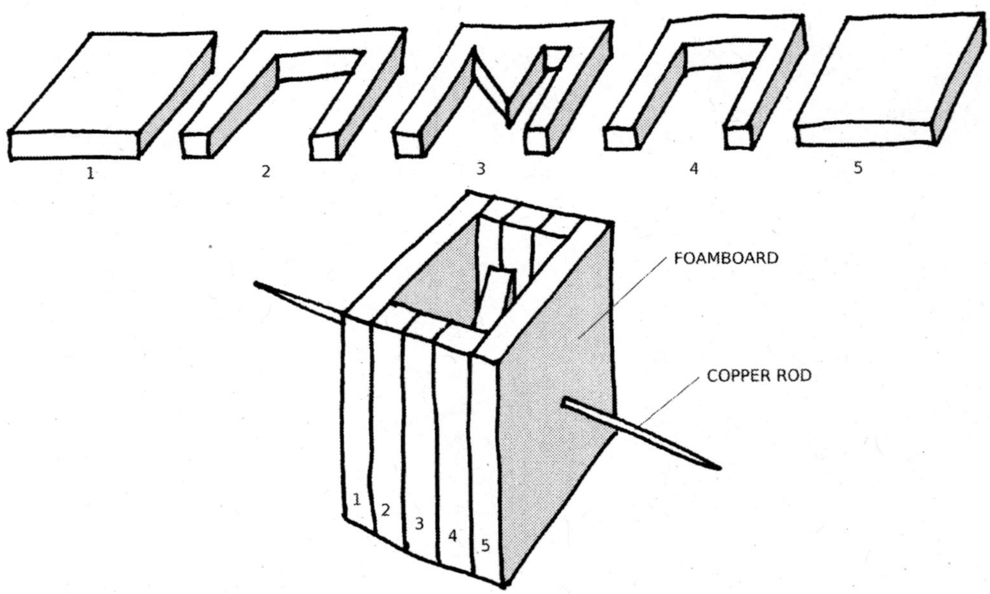

Top: foamboard pieces are cut and stacked. Bottom: assembled waste mould for Slot Box.

This is a one-off mould process, also called a waste mould. The mould is broken to release the concrete. The mix for casting should be wet enough to slump in the mould, to fill corners and spaces, but not so wet that it weakens the end result. The mix I used was a hardware store "sand mix" with added polypropylene fibres (loose) and latex solution (1:3 latex:water). The mix is tamped into the mould using pieces of wood, and tapped lightly on the outside to release air bubbles. (A power palm sander makes a good small vibrator for this scale of work.) Cover the exposed concrete with plastic sheet.

After two days the mould is broken open by releasing the tape. The foamboard that makes the slot (piece #3 in the diagram) has to be broken out.

Bowls

A mould for a bowl form can be made from foamboard and a bowl, as shown below. Any bowl form, such as a mixing bowl, will work as long as it has enough flexibility to be pried from the concrete after it has set (1). A circular cut is made in thick foamboard (you can laminate thinner pieces together with foamboard adhesive to make thick pieces). This is glued or taped onto a base piece (2 and 3). The bowl form is turned upside down inside the cutout circle for the negative part of the form (4).

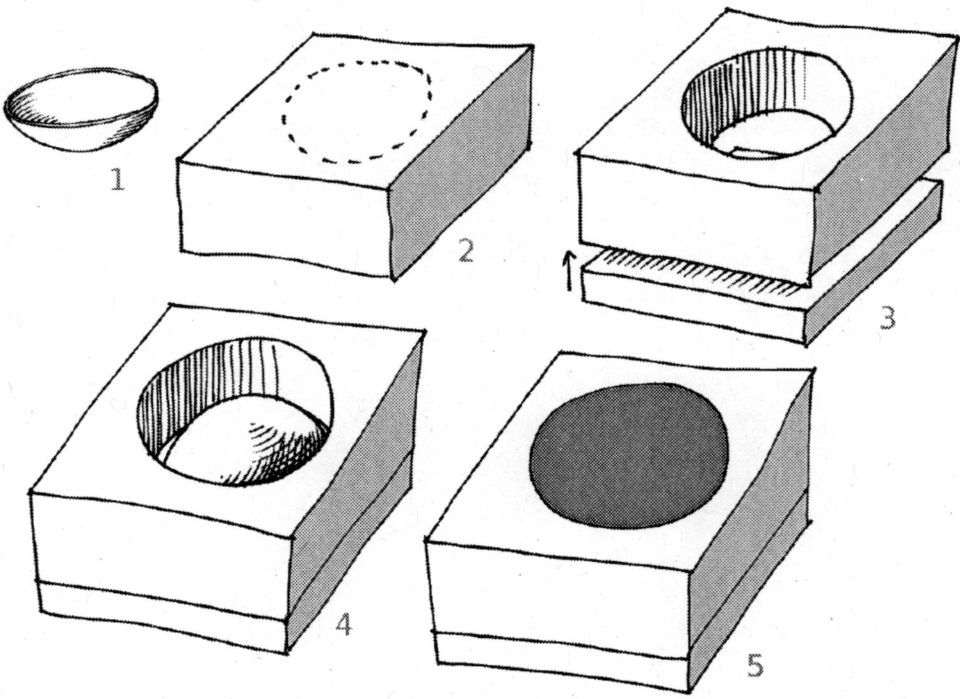

A concrete mix is placed in the mould (5). After it sets the foamboard is broken apart to release the concrete. Then the original bowl is removed. This mould produces a bowl with a flat bottom.

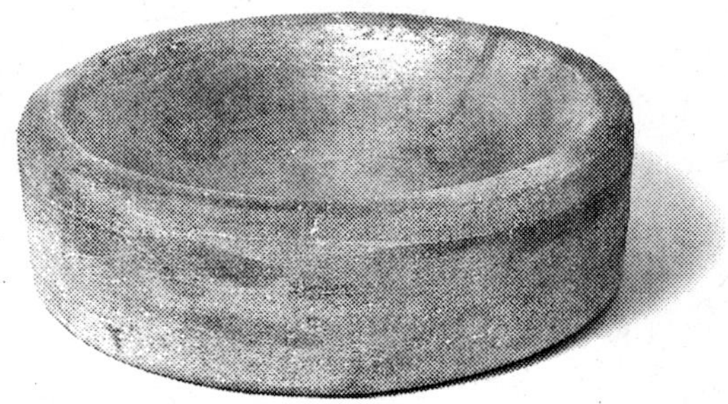

Concrete bowl: cast using the process above. 12" (30 cm) across.

You can also use a bowl form as a one-part mould, and apply the concrete mix to the outside by hand, as was done in the series shown below. A flexible bowl is chosen and inverted on the work surface (1). You might want to cover it with plastic wrap or petroleum jelly as a release agent. A fibre-reinforced mix (glass fibres work well here) is applied starting at the bottom and building it up about 1 cm (1/2") thick, overlapping sections as you go (2). See the **MESH** chapter as this technique is similar to the layer building in that process. When the bowl form is completely covered in concrete (3), cover with plastic and let it set 24 hours. Uncover, make sure the first layer is not dry and add another layer of mix (4). You can shape feet or a rim at this stage.

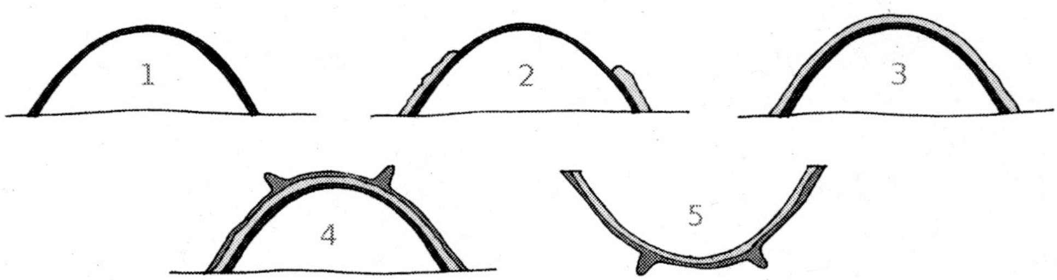

Cross-section series of forming a bowl over a bowl mould.

After another 24 hours, turn the mould and the concrete upside down and carefully remove the original bowl (5). You can shape the rim with a rasp or coarse sandpaper at this stage and add a smooth layer of mix to the inside and outside.

An alternative method for making concrete vessels or containers is to place the concrete mix on the *inside* of the mould, building it up from the bottom. For these thin-walled forms you must use fibre reinforcing in the mix, or embed steel mesh.

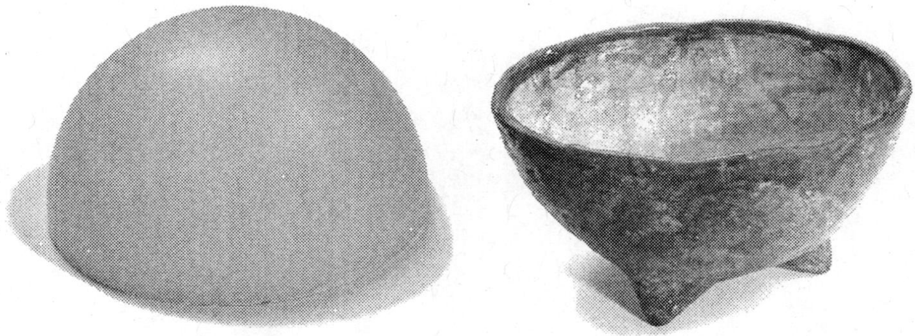

Left, plastic form. Right: three-footed concrete bowl, 13" (32 cm). 1999. A. Goss

The three-footed bowl on the right was made using the technique above, applying a glass fibre concrete mix to the outside of an inverted plastic half-spherical bowl, shown on the left. (This is actually a squirrel guard for a bird feeder.) The initial thin layer applied against the form was a combination of white and grey Portland cements. The second layer was a grey mix of cement, sand and alkali resistant glass fibres. The feet were moulded by hand at this stage. The third layer was pigmented black and rubbed on. Total thickness is about 1 cm (1/2").

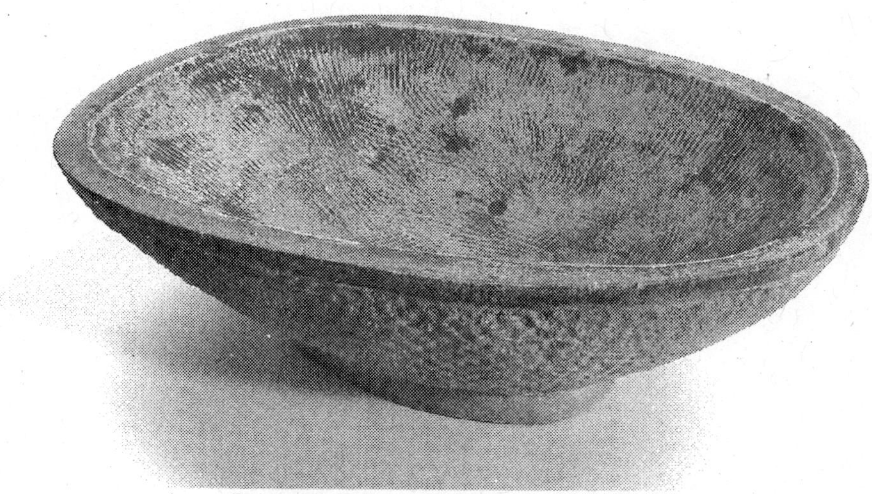

Bowl: concrete, 12" (30 cm), 1999. A. Goss

The bowl above was made of glass reinforced concrete applied by hand over an inverted bowl form also made of concrete. Rubber non-skid material was pressed into the wet concrete to make the outside texture.

Wood and Metal

Moulds can be made out of wood and sheet metal (such as galvanized steel used for roof flashings and stovepipes) either alone or together. The steel gives a smooth planar surface and the wood acts as a support. I use packing tape or duct tape as a way of temporarily holding the mould together. Large moulds would need screws or nails.

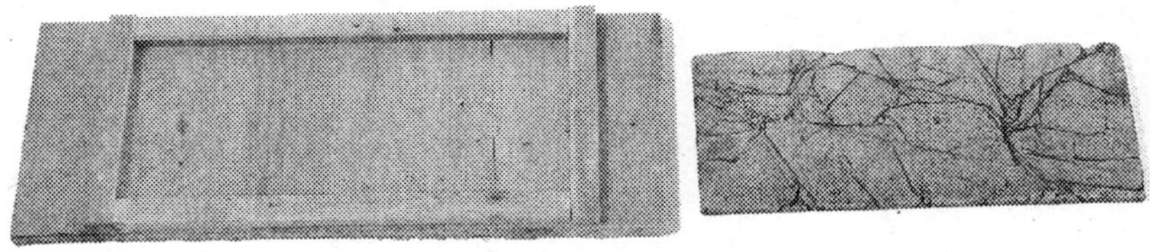

This is a suitable mould for making tiles, made with thin strips of wood and plywood. A cast tile is on the right. The mould was lined with plastic sheet that was wrinkled to give the texture. Surface stains were used to exaggerate the depth of the texture.

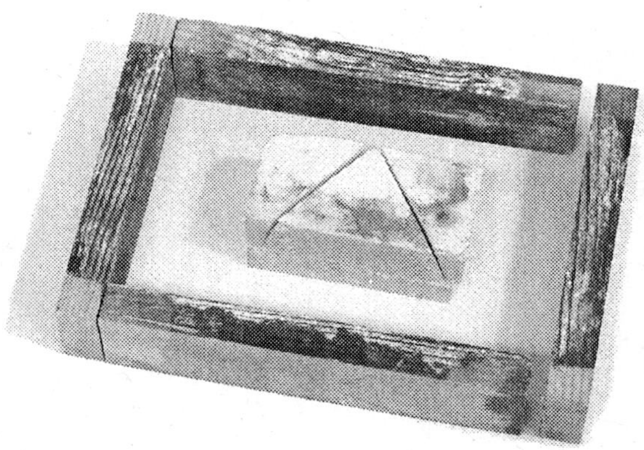

This is a simple small wooden mould used to make the concrete frame piece shown in the **MIXES** section: Feather #2. It was made from 1" X 2" pine, spray-painted to make it waterproof. The wood could also be shellacked. The foamboard pieces in the middle made the negative space, and were cut to come apart without breaking. They also held the silver wire that was embedded in the concrete to hold the pin.

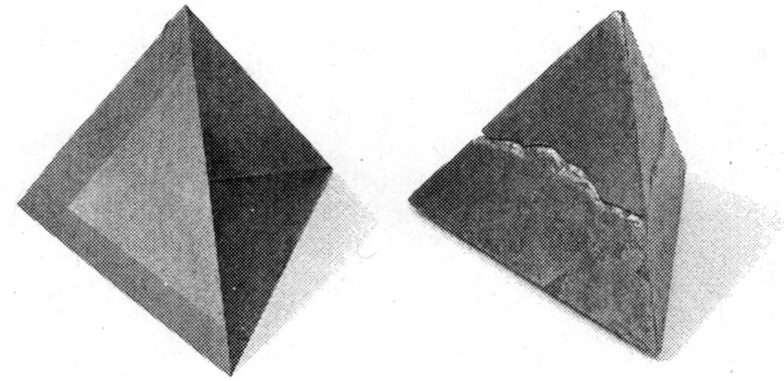

Simple moulds can be made from sheet metal, held together with plastic packing tape. In this concrete box, galvanized steel sheet (left) was used to cast the tetrahedral form. Carved foamboard was used to create the negative space in the box and cut out later. To keep the lid as a separate piece, the box was cast in two stages (see drawings below).

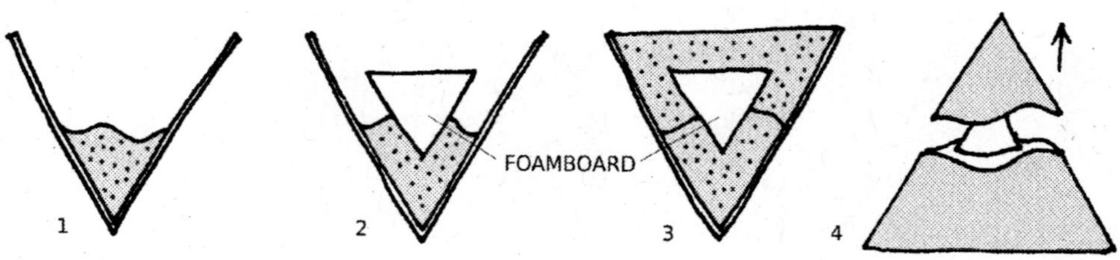

Process used to make the tetrahedral box shown above. 1. Sheet metal mould partially filled with concrete. 2. Foamboard piece to make hollow space is pushed into place. 3. After concrete has set, release agent is applied to exposed concrete and the rest of the concrete is added. 4. Mould is removed, concrete lid is removed, and the foamboard is cut out.

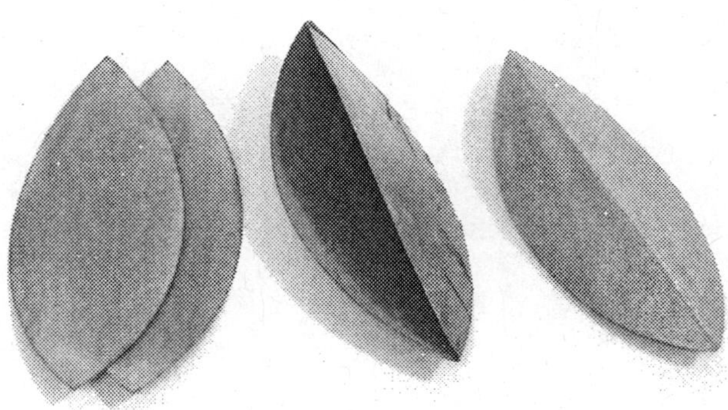

These images shows the same process: left, sheet metal parts for curved form; middle, metal held together with tape; right, the final form. These pieces are about 8" (20 cm) in size. To prevent distortion, this mould and the tetrahedral one above required extra support when casting, such as supporting the mould in sand or a wooden frame.

Miscellaneous

The number of things in which you can cast concrete is limited only by your imagination. Remember that the material has to be strong, waterproof and easy to disassemble, or at least flexible. Think of things that can be nailed or screwed together, things that can be taped or glued together. In producing really small pieces like jewellery, strength may not even matter in the mould material.

I use modelling clay (Plasticene®) to make a negative mould for jewellery-scale pieces. It releases quite well from the concrete, but first make sure that the concrete is hard enough to withstand the stress of pulling it off. A release agent like petroleum jelly helps. You can use sheets of modelling clay to form textures and drawings in combination with wooden moulds to make relief tiles as well.

You can also use pottery clay and cast into it while it is still damp. If you let it dry it will absorb too much water from the concrete. It should be slightly softer than what potters describe as "leather hard." Depending on scale you may need to reinforce the mould. This supporting structure could be made of wood – waterproofed with a sealer of some sort – or plaster. After releasing the concrete you can wash off the clay that sticks to the surface.

Damp sand can be used to quickly make simple moulds, modelling the shape in reverse. The moisture helps it hold together and also stops water in the concrete mix from being wicked out. You can also line the mould with plastic wrap, if you are going to rework or add to the surface of the finished piece.

There are many flexible rubber and plastic mat materials sold in hardware stores. Some have textures that are interesting. But even without texture these materials can make flexible moulds. They can also be used to line other moulds to make them waterproof.

Various plastic forms can be cut apart and taped back together (if necessary) and used as moulds for either casting or forming concrete against the surface. Such shapes as bowls, storage containers and planters can be used. After the concrete has set, remove the tape and the plastic form.

Connections

See the end of the **MESH** chapter for connection solutions.

Mould Releases

See the end of the **OTHER MATERIALS** chapter 5c for mould releases.

10d

Surface Finishing

After you have cast a piece or built on an armature you will probably want to change and finish the surface in some way. This is a summary of ways to alter and refine the surface, both by adding to and taking away from it.

Removing Material

With a partially set surface you can scrape or gouge away concrete using wood, plastic or metal shapes of almost any description: old knives, trowels, metal scraps, coarse wood files and rasps, bristle brushes, old saws or hack saw blades. Because the cement is still active at this stage these tools will have to be cleaned well during and after use or the cement will adhere permanently.

After the cement has set, but before it has cured completely, you can still use coarse files, but I often use very coarse sandpaper (16 to 24 grit, used for refinishing hardwood floors) wrapped around wood to make a disposable "file."

If you use a lot of coarse aggregate in the mix you can remove the surface cement with a brush before a complete set. This will expose the aggregate to give a pebble finish.

WARNING!

At this stage, the partially set cement is still active, alkali and caustic. It can burn your skin. Wear rubber gloves and eye protection. See the **MATERIALS** chapter for more details.

Adding Material

As mentioned before, the sooner you can add a cement-based mixture to the existing surface, the more complete the bonding will be. Any adding should be done within 24 hours of the last layer.

Often a piece will come out of its mould with surface holes or small bubbles. For small holes, say 1 mm (1/16"), mix up a smooth paste of cement and liquid (e.g. latex or acrylic, which increases strength and helps adherence). You can also use metakaolin as an additive, in a ratio of about 1:8, metakaolin:cement.

If the holes are larger, say 3 to 4 mm (3/16"), you have to consider shrinkage, so use stone dust or screened sand. A typical mix might be:

- 1 part stone dust
- 1 part Portland cement
- 1/8 part metakaolin
- latex or acrylic solution

You want this mix to be wet enough to be able to push it into the holes or scratches, but not so wet that it slumps. The drier the mix the less likely you are to have shrinkage problems.

If the cavities you want to fill are larger than 4 mm (3/16") or so, you might want to add polypropylene fibres to reduce shrinkage cracking.

Make sure the concrete surface you are adding to is damp. This means damp, not wet and shiny with excess water – which would combine in the new mix and cause excessive shrinkage. If the surface is dry to the touch it would absorb water from the new mix, so spray the dry surface with water, then wipe off the excess before adding to it.

In any filling of the surface you must think ahead to the desired final finish. You might want to add dye (pigment) to the added mix so that when the surface is sanded smooth, the added material will show as contrast. But if you want the patching to be invisible, you will have to duplicate the colour of your main mix. If the main mix was a stone, sand and cement one, keep the fill mix the same sand/cement ratio. (Hint: think ahead and reserve a small amount of your original dry mix for later hole filling.)

Combinations

This process of subtracting and adding material is a creative one. Tool marks can be filled with contrasting mixtures, and then smoothed to leave some wonderful surface finishes. (See photograph below.) Experimentation is the only way to learn this. A typical process for a simple cast piece might be:

- remove piece from mould after 24 hours
- using a coarse file reshape the surface
- brush away loose particles, rinse off, wipe off excess water
- make a contrasting mix and smooth into surface
- let set 24 hours, covered
- wet sand the surface (180 or 220 grit)
- let set 24 hours, covered

- wet sand with finer grits (320 or 400)
- cover for five more days
- wet sand with 600 grit, rinse, dry
- paint with penetrating sealer
- let dry and wax

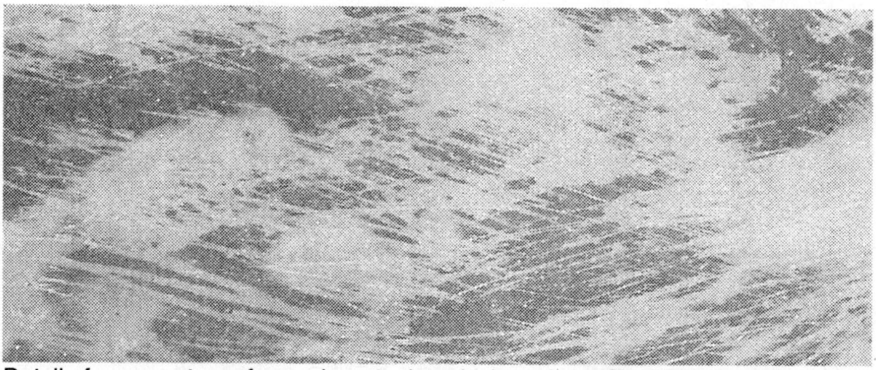

Detail of a concrete surface, where tool marks have been filled in with a lighter mix.

The surface finishing of concrete is a combination of scraping or abrading (either for developing form or smoothing) and filling in negative areas. This process often goes on for several days before you get the desired effect. Remember to keep the piece well covered with plastic between sessions, and also to keep it damp by misting with water while you are working on it.

Surface Colour

Acid stains etch the surface of the concrete, opening it up to penetration. The coloured metallic salts in the stain combine with the hydrated lime (calcium hydroxide) in the surface. The typical colours of black, brown, blues and greens are durable, but usually protected with a sealer (such as acrylic, siloxane, urethane, epoxy). ConcreteNetwork.com has more details. The stains can be applied with a brush or a sprayer – one that will not be affected by acid.

WARNING!

Acid stains are corrosive (hydrochloric acid). Read the MSDS and label carefully. Wear safety goggles, gloves, and a protective apron.

Opaque water-based stains designed for outdoor use also work well but are not as durable. They work best when applied to a cured surface that is damp. Brush it on, and then wipe it off with a damp cloth or sponge, removing most of the dye from the high parts. The dye exaggerates the relief and texture of the surface. If the colour is not strong enough, repeat the process.

To cover concrete completely you can use paint specifically recommended for it. But because of possible colour breakdown due to alkalinity, wait until the concrete is well cured, at least a month.

See **COLOUR AND FINISH** chapter for an explanation of pigments that are integrated into the concrete mix.

Sealing

Various sealers – also described in the **MATERIALS** chapter – can be used after final finishing. Some surface stains may inhibit sealer penetration, like the latex-based opaque outdoor stains. Similarly, if you seal first, the colour may not penetrate. You may have to make a decision about the relative importance of "waterproof-ness" and colour intensity.

Most siloxane or silane sealers are solvent or alcohol based, so the concrete must be dry and cured for it to penetrate well. They should also be applied to a dry, cured surface. Both types of siloxane sealers should then be reapplied five minutes later. This reapplication compensates for areas that may absorb more sealer.

Manufacturers recommend that potassium- or sodium-based sealers should be applied to a surface that is damp, not wet. They chemically combine below the surface with free lime, so the longer they stay wet, the greater the penetration. On a dry surface they do not penetrate well. I usually paint the sealer on twice, and then wrap it in plastic for a few hours. As mentioned previously, these sealers may not be as effective when using pozzolans in the mix.

Both siloxane and potassium sealers are excellent for reducing water penetration in concrete. But even the best concretes are slightly porous and the surface picks up stains easily. If you don't mind the high gloss, wax is a good final solution. I use a paste wax with carnauba. When the piece is dry and clean, rub it in, then buff with a soft cloth. The common acrylic sealers are another possibility. Like wax, they stay near the surface. If applied too heavily they can give an artificial appearance to the surface, making it look as if it's been varnished. Acrylic sealers can also yellow over time.

Summary

The best surface finishes are based on good quality concrete, as dense and non-porous as possible. This results from using the correct proportions in the mix, thorough mixing, minimal air bubbles, and careful tamping into place. Correct curing is important for all the cement to hydrate. If the finished surface is cracked, porous or weak, no amount of sealer will help. But if the surface is good, a sealer can make it better.

11

Ciment Fondu

More information about using ciment fondu can be found in other reference material (see **BOOKS** section). This book is focused on Portland cement concrete. What follows is some introductory information.

Ciment fondu is also called "calcium aluminate cement" or "high alumina cement" (HAC). Lafarge Cement developed it in 1908 as a fast setting cement resistant to sulphate attack, for chimneys and other refractory uses. It's derived from limestone (for the lime) and bauxite (for the alumina). It's often used as a casting medium by artists partly because it develops the same amount of hardness in one day as Portland cement does in 28 days. (A Portland cement-based material will continue to gain strength beyond that time.)

HAC is sometimes mixed with Portland cement to make a quick-setting mortar repair material. This may have applications in the art world.

However, there are some cautions about the long-term strength of ciment fondu. In many countries it is banned from structural or construction uses because it becomes more porous over time. The cement hydrates are not stable over 20°C (68°F) and the porosity allows water penetration that makes it subject to chemical attack and weakening. These conditions would only be a problem for artists making outdoor work exposed to temperature changes and moisture.

A typical method for ciment fondu: a clay original is made, then a two-part plaster mould cast around that. A slurry of ciment fondu about the consistency of cream is painted on the inside surface of the mould (to pick up detail), then a mix of ciment fondu with an aggregate (e.g. sand:cement 3:1) is poured or placed into the plaster mould. After the ciment fondu has set, in about a day, the plaster is broken away.

Ciment fondu sets in about three hours, is hard in seven, and is considered mostly cured in 24. This extremely rapid set time means more heat is given off, and it is critical to apply moisture to the surface as soon as it sets, and for the next 24 hours. This is often done with wet rags.

WARNING!

Ciment fondu (HAC) dust is an irritant to nose, throat, skin and eyes. Wear gloves and a mask when handling it.

12

Gallery

There are many ways of using concrete to create sculpture and objects of functional art. Some artists treat it almost as an industrial material, leaving the finish coarse and matte, like the wall of a contemporary building; others grind and polish the surface like stone. Some use it as a massive, serious material; others treat it with whimsy, surprising the viewer.

I asked six artists to submit images with comments about their own work. The works shown here exhibit a wide range of attitudes and techniques.

Stephen Hogbin

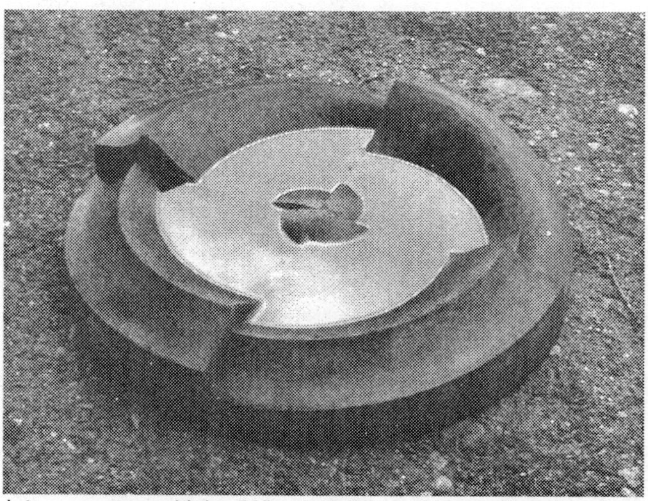

Breaking Quiet: concrete multiple, 23" x 4.5" (59 cm X 12 cm), 2000. Stephen Hogbin.
Rainwater has collected in the form.

Stephen Hogbin is an artist from Charles Lake, near Owen Sound, Ontario, Canada, working primarily in the medium of wood on commissions, individual pieces and installations. He says of *Breaking Quiet* (shown above with rainwater): "The original form is made from wood turned on the lathe. This turned form is cut into four and realigned. A pattern is made from the wood mould in casting plaster. The finished concrete piece is made from the plaster pattern."

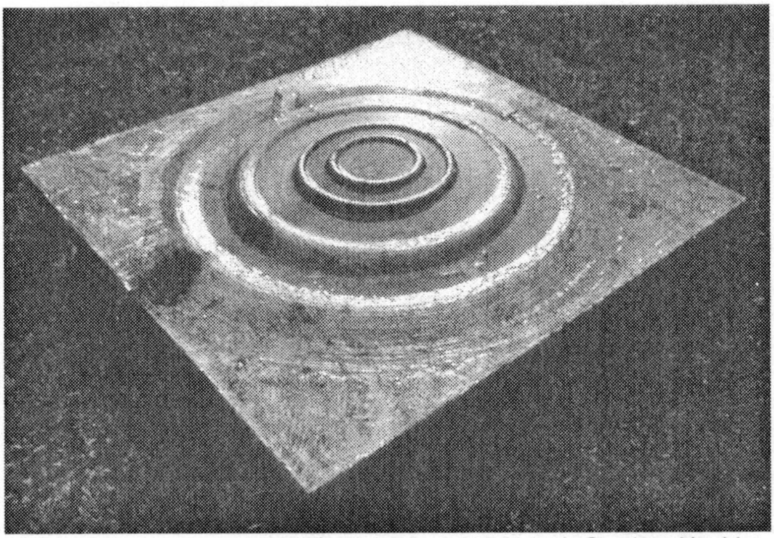

Drumlin Flux: concrete, 48" X 48" (120 cm X 120 cm). Stephen Hogbin.
Shown outdoors with rainwater.

Hogbin also comments, "What has interested me with the concrete forms is the interaction of the form with the environment. One outdoor piece, Drumlin Flux, did the rather extraordinary thing of developing red algae in the shallow outer ring of water and green algae in the centre deeper hemispherical hollow. The leaves in another piece, painted blue, settled in a pattern emphasizing the form while offering the complementary orange/brown colour." For more information about Hogbin's work, visit his website at: http://www.makersgallery.com/Hogbin.

Rachel Reynolds

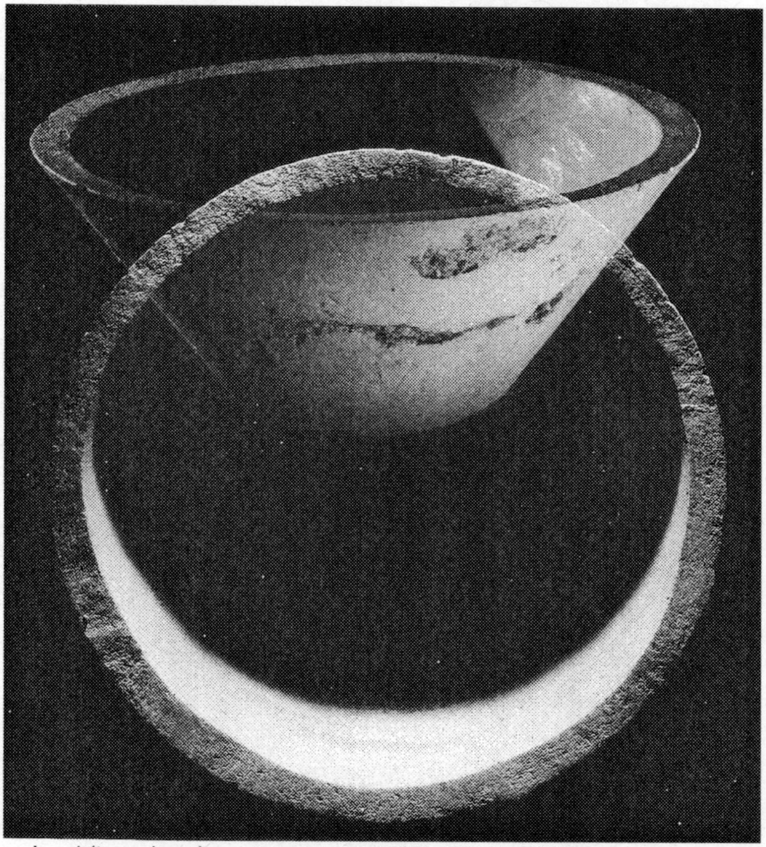

Shallow bowl (two views): cast concrete, 12" X 12" X 7.5" (30 cm X 30 cm X 20 cm).
Rachel Reynolds.

Rachel Reynolds' company, the shed, makes contemporary garden furniture in London, England. Commenting on her business Reynolds says: "The desire for a modern, elegant garden can be frustrated by the lack of contemporary furnishings. At the shed we are committed to creating high quality, hand made garden features and fittings with a contemporary feel using concrete as an alternative to terracotta.

"The shed creates individually crafted pieces from concrete which has an air of durability and permanence. We make large bowls and a variety of benches and stools using a variety of methods including casting and hand building forms around a mesh frame. As each piece is individually hand crafted in our studios each is completely unique.

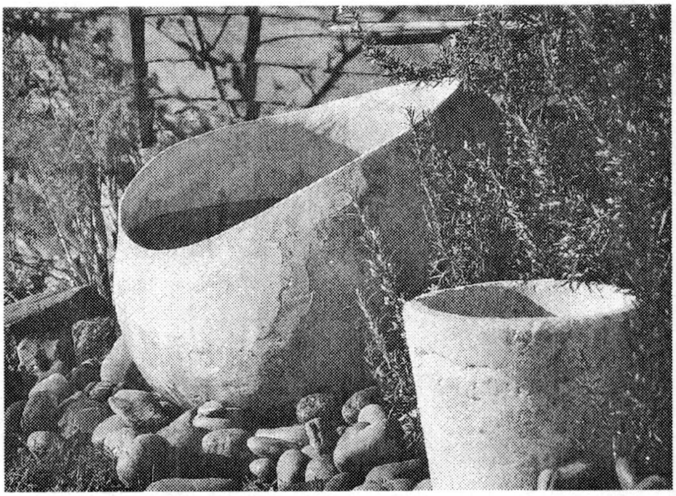

Left: spherical vessel, hand formed around mesh frame, 26" X 18" (66 cm X 45 cm).
Right: round pot, cast concrete, 12" X 12" (30 cm X 30 cm). Rachel Reynolds.

"My initial motivation for working in concrete was the immediacy it allowed when creating a piece: the ability to create permanent structures, which were not models or maquettes without the need for expensive equipment. Relatively large pieces can be made using just expanded mesh, wire, tape, a pair of pliers, a trowel and a bucket. Once the initial love affair was over I soon discovered that although the actual equipment was extremely low tech, the material that I had chosen to work with was technically challenging. When you are hand building large pieces it is sometimes impossible to know if it is adequately reinforced. So the correct strength of the concrete is essential.

"My favoured approach is to use a high ratio of glass fibre/aggregate (usually washed sand) to cement for the base coat with a low water content, with added latex for workability. Then I use a higher ratio of cement to sand for the topcoat with again, a low water content to ensure strength. Another factor which must be considered when building pieces is the time it takes to actually hand trowel the concrete onto the frame and how much the ambient temperature affects the time you have before the concrete goes off. So it is important to limit the area that you work on to fit the time you have available."

Nicholas Rees

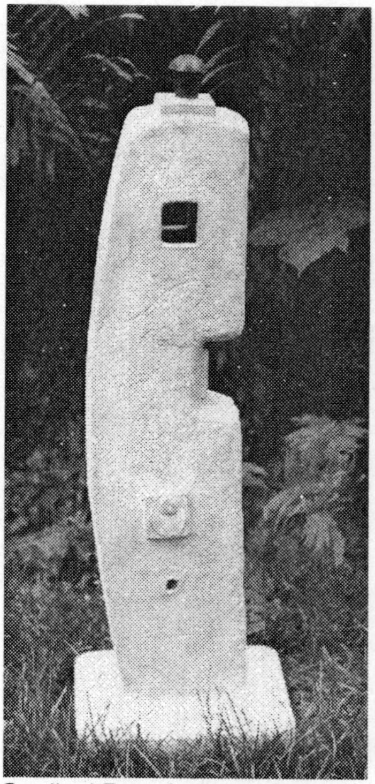 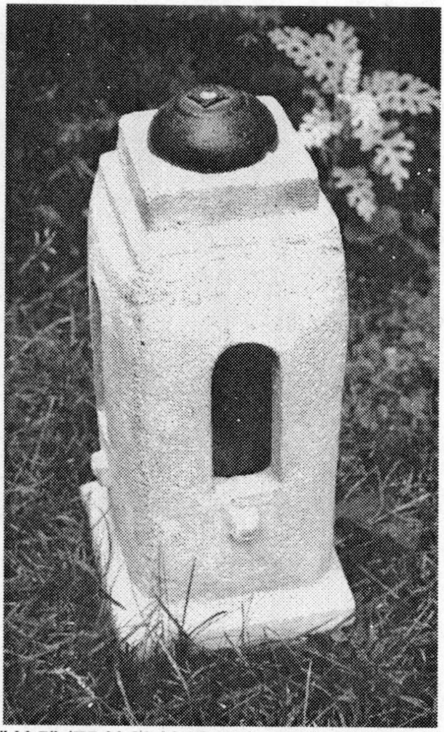

Left, Cardinal Post: ferro-cement, 30" X 8" X 5" (75 X 20 X 12 cm), 1995. Nicholas Rees.
Right, Carillon: ferro-cement, steel, 14" X 7" X 7" (35 X 17 X 17 cm), 1998. Nicholas Rees.

Nicholas Rees is an artist from Kitchener, Ontario, Canada. He makes large outdoor pieces sometimes incorporating found iron objects. He started working in ferro-cement in 1986, working with chicken wire and a bag of sand mix cement. "The first piece was a four foot tall tower intended for the garden, with an arched roof. It has survived nearly 17 winters and seems set for many more," Rees says. He still keeps the technology to a minimum, using sand mix, chicken wire, pliers, snips and a small cement trowel.

Rees comments: "My own work is concerned primarily with reclaiming and re-sanctifying space; space that has been appropriated by first the industrial, and now the post-industrial age. In a garden setting these sculptures provide a tranquil space and often seem to site themselves.

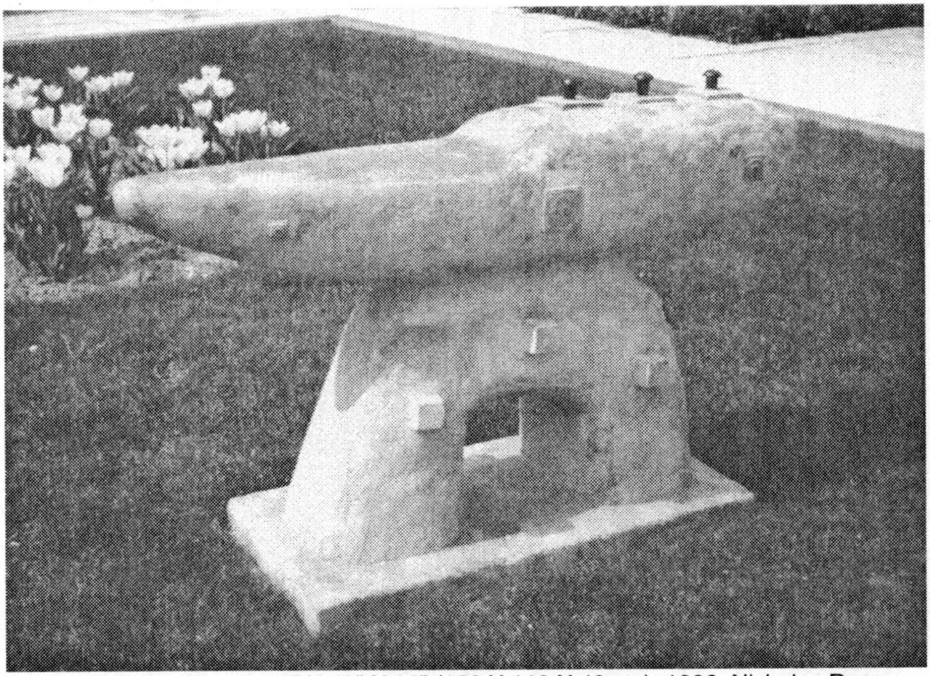

'Anvil', ferro-cement. 66" X 43" X 16" (168 X 110 X 40 cm), 1996. Nicholas Rees. Collection, City of Kitchener

"Ferro-cement is exceptionally versatile, not only in its ability to be shaped, but also the way it combines with other materials, particularly wood, copper, steel, stained glass and tile. I have found it to be a remarkable material, and 16 years on still feel that I am only just beginning to learn something about it."

Ilene Dunn

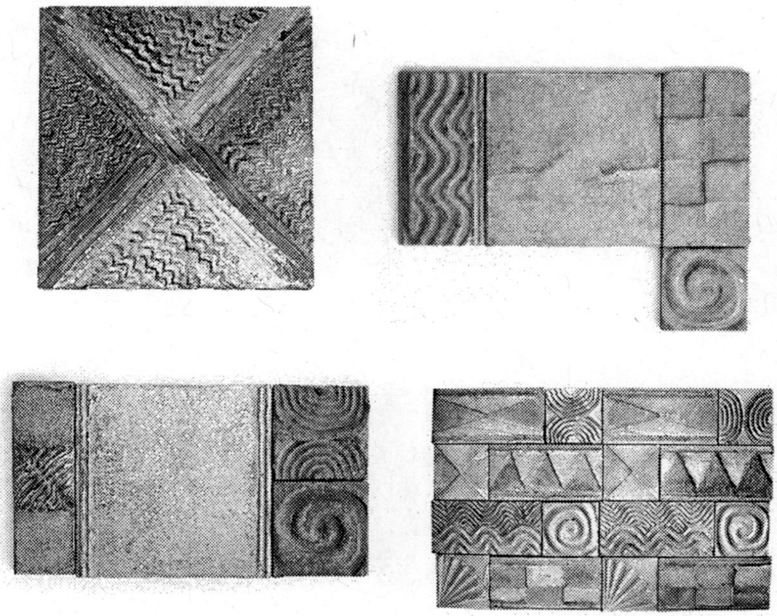

Concrete tiles: various sizes, 3" X 3", 2" X 6", 3" X 6" and 6" X 6" (7.5 X 7.5 cm, 5 X 15 cm, 7.5 X 15 and 15 X 15 cm). Ilene Dunn.

Ilene Dunn of Santa Fe, New Mexico, uses cement with colour to produce sculptural reliefs. She has been creating textured surface art and design for over 15 years. These images show groups of tiles. Dunn explains that she makes these tiles using grout/cement on top of quarry tiles: "I used a base pouring of black/brown grout (drawing the design on this layer), then I applied by brush a wash of coloured mortar(s) diluted with admixture (acrylic additive) – using a spray gun or spray bottle to bring up the under design. To further enrich the surface colour, I added a third coat of either another customized grout colour – meaning, I added coloured pigments to already existing grout colours – or a spray of acrylic paint wash (rust for instance)." They have a water-based penetrating sealer.

What Dunn likes about cement is its "ancient stone quality." She adds: "The creative properties of cement have never ceased to captivate me. I continue to make new discoveries and to be inspired by the new avenues of expression and creativity that it presents to me. If I were to say what this medium of expression holds for me, I would say that it is about the integration of spirit and matter – human spirit expressing and moving through solid mass." For more of Dunn's work, see her web site at http://home.earthlink.net/~idunn/.

Bill Loney

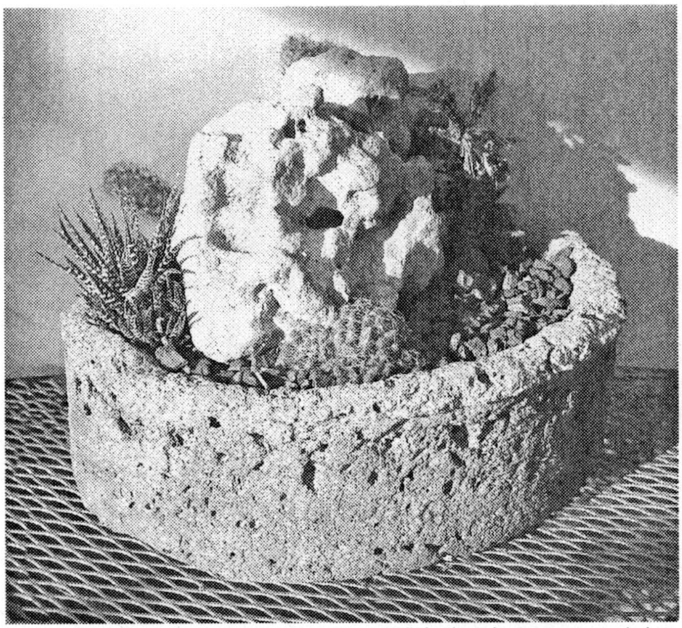
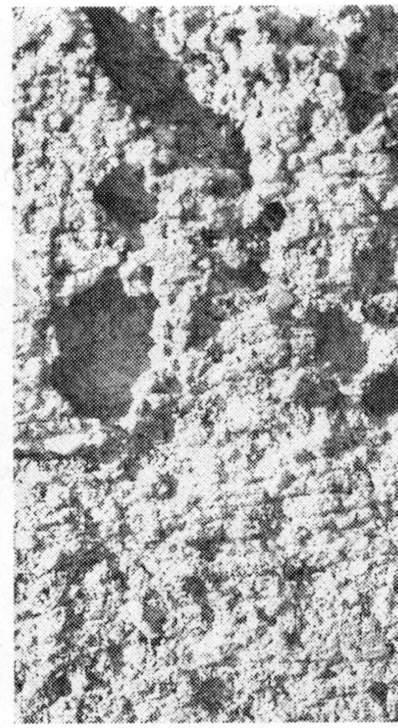

Hyper tufa planter, about 18" (46 cm) across, and detail of surface. Bill Loney.

Bill Loney and his wife Dawn have designed and built a large garden in Big Bay, Ontario, called Keppel Croft Farm and Gardens. To make planters for alpine plants Loney uses a hyper tufa mix of:

- 1 part Portland cement,
- 1 part sand and
- 1 part of any of the following:
    - peat moss, sphagnum moss, vermiculite or perlite,
    - or any combination of moss, vermiculite or perlite
- water

The mix is packed into a container-shaped mould lined with plastic, let set for 24 to 48 hours, then removed and wire brushed to emphasize the texture. The moss, vermiculite and perlite make the concrete lighter in weight than standard concrete. Plastic film canisters are used to make the drainage holes in the mould. Hardware cloth (steel mesh) is used as reinforcing.

Loney likes to use found objects and materials that he discovers as moulds for concrete objects displayed throughout the garden. In the lantern below, sections of concrete were cast in moulds made from such materials as wood trim, a metal lampshade and pieces of foamboard. He recommends using such things as the glass balls used for old lamps and plastic dessert moulds from the kitchen. Loney builds some hyper tufa pieces using a pile of damp sand as the basic shape around which the concrete mix is packed. He also notes that using plastic wrap or sheet can leave a very smooth glossy surface to a cast concrete piece if it is left in the mould for several days.

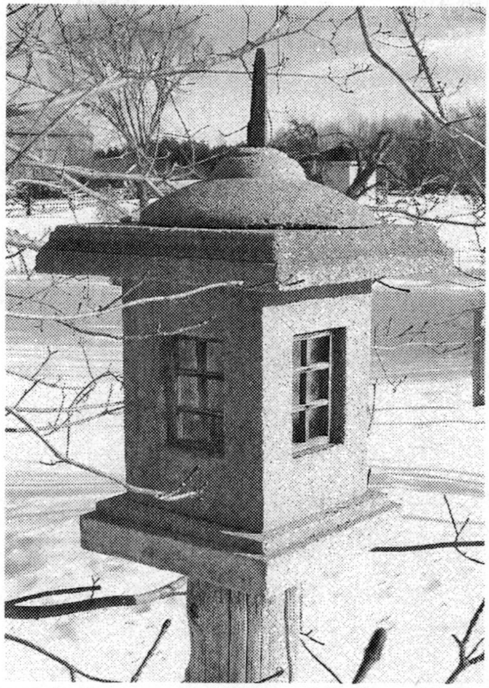

Lantern: concrete, copper, wood, about 24" (60 cm) high. Bill Loney.

Joaquin Aviña

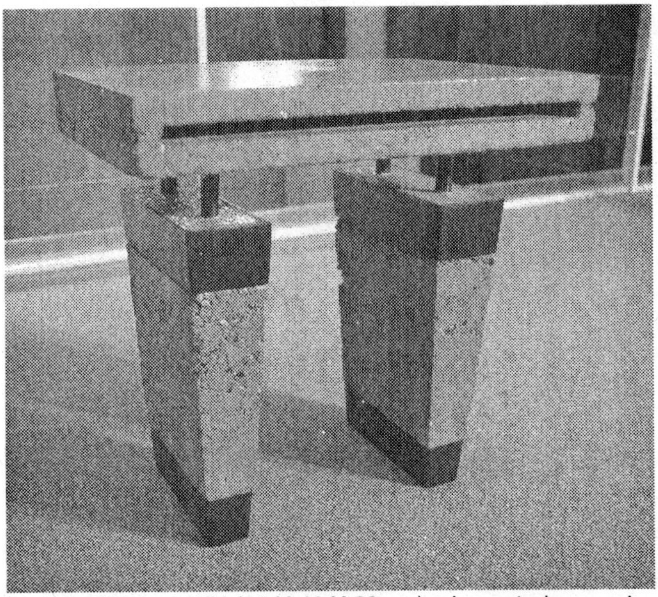

End Table: 24" L x 18" W x 25" H (61 X 46 X 63 cm), pigmented concrete, steel, 2001.
Joaquin Aviña.

Joaquin Aviña is a sculptor from Minneapolis, Minnesota, who has been working with concrete to make furniture. In the End Table above he used a mix of Portland cement, 3/4" coarse aggregate, sand and water. Aviña explains: "The top piece was cast in a mould that created a hollow middle, of 3/4", from front to back. The thickness around is uniform at 1 1/4". Four pieces of steel were welded together for both the top and the bottom bands of the legs, and then the corners were sanded round. Once these pieces were completed, I tied them together with a steel pole in the center of each leg. Each piece tapers towards the floor and a form was created around the steel skeleton. I mixed a red dye into the wet mix and poured that into each form. I wanted to leave the concrete as is to emphasis the juxtaposition of this material in one's home and also to give it a unique texture and look. After the piece cured, I rolled on a sealant to protect the surface."

Whenever Aviña talks about his interest in concrete, he usually gets one of two responses: "The blank stare as they try to fathom the idea of their sidewalk or driveway as a piece of furniture or the verbal, 'isn't that heavy?' Indeed, both these reactions are what drive me to continue, and the fact that I love contemporary urban decay. Why not look at non-conventional materials as useful media in art and design? Concrete has been used as far back as the Greeks, if not farther, but somewhere along the way we have just looked at concrete as a building material. In recent years cement has come into the public eye as more than the material for patios and driveways. You can see its use in sculpture, interior design and furniture design, to name a few examples. And furthermore, why not heavy, or the appearance of heavy? Perception plays greatly on our psyche, and often times influences our reactions. But more common is the reaction of amazement when someone takes an unusual non-traditional material such as concrete, and pushes it in new directions and achieves great success."

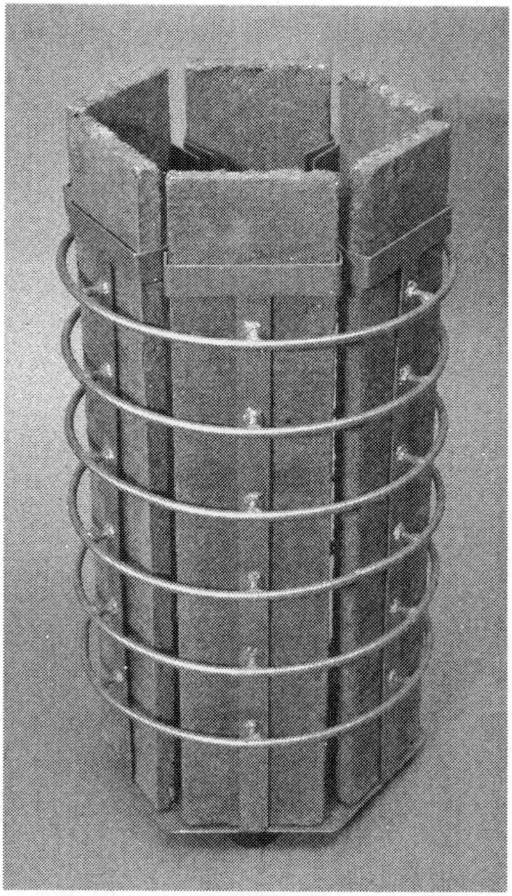

Floor Lamp: concrete, steel, 36" H x 18 " W (91 X 46 cm), 2001.
Joaquin Aviña.

Aviña describes the floor lamp: "This is a concept piece. What would a concrete lampshade do to a lamp? Darken the room. So instead I created something that controlled the direction and intensity of the light. A steel exterior frame was created and six armatures were added to hold each concrete sleeve. These were cast 3/4" thick and 6" wide. I added a stain to the outside surface. A light fixture was added to the inside and a long light bulb was used. Finally, three 2" black rubber stoppers were attached to the bottom for balance."

13

Environmental Concerns

It is impossible to write about concrete without at least mentioning some environmental concerns. You should be aware of the environmental impacts of this material, both positive and negative, so that you can make informed decisions about its use.

Negative Impact

The big issue is the pollution produced in its manufacture. Carbon dioxide (greenhouse gas) is produced by fossil fuels often used to fire the kilns: one half ton of CO_2 for every ton of concrete. Carbon dioxide is also produced as part of the chemical change: one-half ton of CO_2 is produced for every ton of Portland cement produced. That's a total of one ton of CO_2 for every ton of Portland cement. Some estimates say one-eighth of the world's greenhouse gases are generated in the production of cement. (Source: *Alternative Construction – Contemporary Natural Building Materials*, ed: Lynne Elizabeth and Cassandra Adams, John Wiley & Sons, 2000.) The production of steel, used as reinforcement, also uses large amounts of energy. But, to put things into perspective, the average car generates over 5 tons of carbon dioxide a year.

The raw materials used to produce cement and concrete – limestone, chalk, shale, clay, gypsum, stone aggregates – are non-renewable, although they are not scarce.

Some concretes are poorly made and don't last. They are porous or made with too much water. These concretes become susceptible to freeze-thaw breakdowns, or penetration by acid rain. Demolished concrete becomes a disposal problem as a solid waste. You have a responsibility to make a product that will last a long time.

Positive Ideas

There is a good side. Concrete, if made properly, does last a long time. The Pantheon built in Rome over 2,000 years ago is still standing. Finished concrete is basically inert, so is not an environmental problem until it is demolished when it becomes a landfill problem.

You can lessen the proportion of cement in a concrete mix by using more pozzolans, particularly ones like fly ash (a waste product from coal-fired generating stations), blast-furnace slag, silica fume (a waste product from the manufacture of silicon or ferro-silican metals) or metakaolin (made from kaolin, a type of clay). Pozzolans actually improve concrete's qualities. See **MATERIALS** chapter.

You can incorporate other recycled products in the mix as aggregates: old concrete, broken bricks. Recycled glass can be a problem; some types of glass are okay, while some contribute to concrete breakdown because the silica in the glass is reacting with the free lime. New research is indicating that adding pozzolans (thereby reducing the free lime) may allow more recycled glass to be used as an aggregate. (Source www.engelhard.com)

14

Troubleshooting

What's gone wrong and can it be fixed now or next time? This is a brief description of some common problems with explanations of what might have happened.

Mix too wet

You're mixing the concrete and all of a sudden it's like soup. You may be tempted to use it for casting, but don't. It will be too weak. Add more dry mix in the same proportion as the original dry mix until it is the right consistency.

Mix too dry

This is preferable. Add more liquid, small amounts at a time, while mixing.

Mix is setting too fast

You may have mixed a batch of concrete for applying to an armature and after five minutes it's starting to get so hard you can't push it in place. This is probably a "false set." Try remixing. If that doesn't work add a tiny amount of liquid and remix. Is it still too stiff? Discard it. You probably have a fast initial set caused by high temperature. Next time use a colder liquid, or work in a cooler location.

Sometimes if your mixing containers are dirty with old cement, the old cement will act like a catalyst and speed up the setting time.

If your mix is setting up after 30 to 60 minutes, this is what is supposed to be happening and it's probably too late to do anything with it. If you remix it at this stage you will weaken it. Next time you will have to make smaller batches, work faster, use colder water, or use a chemical retardant.

Mix is setting too slowly

This is most likely due to a cool location or using cold water. If the object you are working on is portable, carry it into a warmer location. Next time use a warmer location, use warm water, or be patient. The cement you are using may also be too old. Old cement absorbs water from the air, starts to hydrate, and loses its strength.

If the surface of concrete in contact with a metal mould surface is not setting, this can sometimes be a chemical reaction with copper alloys in the metal, slowing down the set time. Spraying a release agent on the mould, or coating it in plastic film or packing tape, will prevent this. If the concrete is in contact with an unpainted wood surface in a mould, water can be pulled out of the mix so the cement does not cure properly.

Air bubbles on the surface

Some air bubbles are unavoidable when casting. Air gets trapped during the mixing. Vibration helps to float the bubbles out. Tap the outside of the form after filling the mould. I sometimes use en electric engraver to vibrate small jewellery moulds. A palm sander can be used for castings up to 24" or so. For very large cast-in-place work you can rent a concrete vibrator. Caution: too much vibration will separate the aggregate from the cement, so be careful. When casting into some mould materials (such as silicone rubber), spraying on a surfactant (such as a dilute solution of dish detergent and water), is recommended.

Cracking

The first thing to consider if the concrete develops cracks is that you used too much water. This leads to excessive shrinkage. The second thing to consider is the quantity of aggregate. If you don't use enough aggregate, or if the concrete is too rich with cement, you will have excessive shrinkage. Remember that aggregates don't shrink, but cement paste does.

Cracks will sometimes develop where you use one type of mix on top of another. A smoother mix may have less aggregate in it and shrink more than the layer underneath, creating hairline cracks.

Surface cracks can also develop where the cured concrete is exposed to severe temperature changes, and is too thick. See the **STRESSES** chapter.

Cracks that are deep and develop long after the concrete has cured could be due to insufficient or improperly placed reinforcing. Stresses will crack concrete over time, unless the reinforcing is appropriate. Make sure the rebar surface is sufficiently rough.

Frost Cracking

Frost cracking can be prevented first of all by using air entrainers. Microscopic bubbles in the air-entrained concrete allow moisture to expand and contract in freeze-thaw cycles without cracking the concrete. Air entrainers only work when using power mixers – simple hand mixing does not create the air bubbles necessary. Secondly, high quality concrete can prevent many freezing problems. This means high density, impermeable, strong concrete, by using such additives as pozzolans. It also means using as little water as possible by adding superplasticizers (water-reducing agents) as mentioned in the **MATERIALS** chapter.

Flaking off

Porous concrete, or concrete not properly sealed, will flake off in areas that receive frost. Moisture gets into the surface, freezes, expands, and forces concrete off the surface. Concrete made with too much water is porous. Some aggregates are porous and trap water; when the water in the aggregate freezes, it expands and causes flaking off.

If layers in a built-up armature construction split apart there are several possibilities: you waited too long to add the final layer and it hasn't properly bonded; the mixes are too different and one has shrunk more than the other. The solution is to apply the next layer within a day or two of the original one having set. Try using a thin slurry painted on between the layers (a mix of cement and water, or cement and latex) and add the next layer while the slurry is still wet. If adding a layer to older concrete, paint on a thin layer of latex first, then apply the layer of new concrete.

Reinforcing metal shows through

Sometimes when you're grinding down the final surface, a piece of reinforcing steel shows through. On an outdoor piece it will rust and stain the concrete. Chip away the concrete around the steel, enough so that you can nip off the end of the wire below the surface with wire cutters, and patch the hole with the same mix used for the main piece. Alternately you can grind the end of the wire below the surface. Fill the hole with a cement-sand mix slightly above the surface; then when it sets, sand it level.

Appendix

a. Sources of Supply
b. Books
c. Links on the Internet
d. MSDS Portland Cement
e. Glossary

a

Sources of Supply

Expect to be frustrated in your search for some of these materials. The concrete supply business is not set up for artists. You may discover you have to buy a truckload. It would be impossible to list every supplier of concrete-related materials. This is general, generic information about where to find supplies, with some specifics for hard-to-find materials. *A company specifically mentioned is not meant as an endorsement*; chances are it's the first or only source I could find. Those companies listed are mainly Canada and U.S. based. See the **MATERIALS** chapter for a description of some of the additives. Many suppliers and manufacturers have web sites: check the **LINKS** section for these and other specific sources through the Internet. This information may become out of date; consider this list a starting point.

Thanks go to a lot of people for emailing me with sources of supply in their region. I encourage readers to continue to do that so that I can keep the web site up to date.
email: concrete@andrewgoss.com website: www.makersgallery.com/concrete/

Portland Cement

Widely available at local lumber companies, hardware stores and construction supply outlets. Type I is the one for general-purpose use, and will probably be the one you find. White Portland cement may be more difficult to find and is more expensive. Try masonry supply outlets for the white Portland because it is used to make white mortar.

High Alumina Cement (Ciment Fondu)

Sculpture supply stores. In Canada, Webster & Sons, Montreal, QC, and Mississauga, ON.

Aggregates

- Crushed stone or gravel can be found at local ready-mix concrete suppliers, masonry suppliers, or landscape suppliers.
- Sand can be found at these places too, and hardware stores sell it in bags. Sharp sand (also called brick sand or mortar sand) is best.
- Vermiculite: garden suppliers, nurseries, hardware stores, building suppliers.
- Perlite: garden suppliers, nurseries, hardware stores, building suppliers.
- Styrofoam or polystyrene beads: try B-Lite, www.b-lite.com/bsttext.htm.
- Glass spheres: Douglas & Sturgess, San Francisco, CA. www.artstuf.com

Other Materials

Stone dust (limestone or marble): try local quarries. Limestone (dolomite) is also sold in gardening-related places because it is used to treat acidic soil. Marble dust is sometimes sold as "swimming pool aggregate," so try swimming pool suppliers or contractors. Also check pottery supply outlets.

Latex or Acrylic Additive

Latex additive is usually available in four-litre or one-gallon containers, from masonry suppliers or construction supply outlets (such as contractor equipment rentals). It looks like milk and has the consistency of cream. One brand is Concentrated Latex Adhesive made by Construction Product Distribution Services, Concord, ON, Canada 905-669-5013. Duomatrix-C is another polymer additive (latex) available through www.smooth-on.com.

Some building suppliers and masonry suppliers have acrylic admixture, either as a pure liquid or mixed with latex. Sometimes it is described as a concrete bonding agent. One brand is Homecrete Acrylic Cement Bond, from Home Hardware in Canada. Home Depot in Canada and the USA carries another brand

Fibres

Plastic

Polypropylene fibres (e.g. brand names Fibermesh® or Fiber Ad®) can be difficult to find in small quantities. Try your local ready-mix concrete supplier. It's usually supplied in 1 lb. bags, enough for a cubic yard of concrete. The fibres are about 1 cm (1/2") long. Sometimes it's called "micromesh." Fibermesh® is in Tennessee at 800-635-2308, 423-892-8080. Fiber Ad® is in Charlotte, North Carolina, at 704-398-0954 (**LINKS**). Form and Build in S. Ontario carries fibres.
PVA (polyvinyl alcohol) or Nylon fibres: Nycon, Inc.,101 Cross Street, Westerly, RI, 401-596-3955, 800-456-9266 nycon.com (**LINKS**).

Glass Fibres

A reinforcing additive, they must be alkali resistant, usually written as "AR." The chopped strands are probably the most useful. In North America the chopped strands and chopped strand mat is available from Ball Consulting, Ambridge, PA, 800-225-2673, 724-266-1502, fax 724-266-1504 or their branch in Tempe, AZ (**LINKS**).

Mesh

Hardware cloth, expanded galvanized mesh and chicken wire can usually be found at lumber stores, masonry suppliers or hardware stores. The expanded galvanized mesh is used as a base for exterior stucco on houses and sometimes called stucco lath. For the wire to hold it together, ask for binding wire, stovepipe wire or galvanized wire, usually available at hardware stores.

Steel

Thin steel rod for reinforcing can be found at local steel suppliers. Rebar made specifically for concrete construction can be found at lumber or construction supplies outlets. The most common sizes are #3 (3/8"), #4 (1/2"), and #5 (5/8"). For small-scale work you can also use coat hanger wire or galvanized wire used for fencing repairs.

Air Entrainers and Plasticizers

These are difficult to source. Master Builders manufactures chemical admixtures, but I haven't been able to buy small quantities directly from them. Their address is: 23700 Chargrin Blvd., Cleveland, Ohio, 800-MBT-9990 (**LINKS**). Ask your local ready-mix supplier if they will sell you small quantities or try a masonry supplier. Also try Grace Construction Products, a specialty chemical supplier, Cambridge, MA, 617-876-1400, 877-423-6491 (**LINKS**). Specco Industries is another manufacturer of a variety of admixtures, including melamine plasticizer: 13087 Main St., Lemont, IL 60439, 630-257-5060 (**LINKS**). Or Euclid Chemical, www.euclidchemical.com. Air entrainers are only essential for outdoor work in frost zones. Plasticizers (water reducers) are not essential either, but they can increase strength by allowing you to use less water. If you can't find a plasticizer like Pozzolith® or Polyhede®, you can use a latex or acrylic additive instead of water, although it is more expensive in large quantities.

Silica Fume

An extremely fine dark powder, silica fume is a waste product from the smelting of metals. As a concrete additive it increases strength and reduces permeability. Available through Master Builders and other suppliers, it is difficult to find in small quantities. Some cement manufacturers add silica fume to their SF blended cements.

Metakaolin

One brand made by Engelhard is called MetaMax® (**LINKS**) and is a fine white powder with similar qualities to silica fume. It increases strength, decreases permeability, and lessens shrinkage. It is used as 5 to 15% by weight of the Portland cement. It may be difficult to find in less than 50 pound bags. A Canadian source for this is Stochem at 905-458-6888. A U.S. source: www.ConcreteCountertopInstitute.com for 40 lb bags. Try suppliers to pool contractors. (**LINKS**)

Mould Supplies

Art and sculpture supply stores are the best places to find mould-making supplies. Try Sculpture Supply of Canada, 345 Munster Ave., Etobicoke, ON, 416-234-1015. Artcast in Georgetown, ON, 905-877-5455. Webster & Sons, Montreal, QC, and Mississauga, ON. In the U.S., try The Compleat Sculptor. Polytek.com in Easton, PA has an online store. Douglas & Sturgess, San Francisco, CA. www.artstuf.com. See **LINKS** section for more information on these suppliers.

Sealers

Some sealers for waterproofing concrete can be found in your local hardware or lumber store. Others can be obtained through concrete or masonry suppliers, or directly from the manufacturers.

Acrylic

Most of the readily available ones are acrylic-based and sit close to the surface. They are available in paint, hardware and lumber stores.

Siloxane

Silane or siloxane sealers combine chemically inside the concrete leaving a natural surface, but may be difficult to find. Check with concrete and masonry suppliers. Try Lighthouse Products for their siloxane sealers, PO Box 1253, New Smyrna Beach, Florida 32170, 904-428-8888 or 800-228-5537. In Canada, Gem-Gard SX®, a siloxane sealer, is available in large 20L pails from the distributor Form and Build (**LINKS**). Smaller quantities may be available from CPD (see latex above), through a local distributor.

Potassium and sodium silicates
V-Seal, a clear, colourless and water-based sealer, is available through V-Seal's web site in North America (**LINKS**). There are other potassium-based alternatives to this sealer such as Seal-It® in Saskatoon, Saskatchewan, Canada and Bio Vee® from Progressive Epoxy Polymers, 48 Wildwood Dr., Pittsfield, NH, 603-435-7199 (**LINKS**). Also GroundZero from www.primeshop.com.

Dyes and Pigments

Integral dyes

As powders or liquids that are mixed into concrete, dyes or pigments can be found in local hardware or lumber stores, or try companies that manufacture concrete block. One Canadian supplier of "mortar colour" is Interstar, 139 Rhéaume S., St. Elie d'Orford, Québec, J0B 2S0 Canada, 819-563-9975. U.S.A. suppliers: Solomon Colors, Springfield, IL, 800-846-2599, Arizona Oxides, El Mirage, AZ, 800-576-1500 (**LINKS**) and Compleat Sculptor (**LINKS**) in New York.

Surface dyes

I've had good results with acrylic-based opaque stains that are used for outdoor wood, but I don't know about long-term breakdown of colour. These are easily available in paint and hardware stores. Special dyes or stains are also made for rubbing into the surface of cured concrete and can be obtained through craft supply stores or hardware stores. One brand is DecoArt's "Patio Paint."

Acid stains

Acid stains etch the surface of the concrete, thereby opening it up to penetration. Try concrete and masonry suppliers for these stains. U.S.A. suppliers are Kemiko Stains, 903-587-3708 (**LINKS**), Triple-S Chemical, Los Angeles, CA, 800-862-5958 or 323-261-7301 (**LINKS**) and NeGeo Products, Salt Lake City, UT, 801-388-2899 (**LINKS**).

Safety Equipment

Most of the things you need to protect yourself, such as eye protection, respirators, and different kinds of waterproof gloves, are available at safety equipment and safety clothing suppliers. You can get some of these things at hardware and building suppliers, but if you need specialized equipment or advice, I would recommend the safety equipment suppliers. Latex or vinyl gloves are sold in drug stores.

b

Books

These are a few books that may be helpful using concrete and cement as an art material. I have added some comments. Some of the older ones may be difficult to find or may be out of print.

Abercrombie, Stanley. ***Ferro-cement: Building with Cement, Sand, and Wire Mesh***. New York: Shocken, 1977. This book is mostly about architecture, but includes boats, basic principles, and methods.

Andrews, Oliver. ***Living Materials: A Sculptor's Handbook***. Berkeley: University of California Press, 1988.

Cheng, Fu-Tung. ***Concrete Countertops***. Newton, CT: Taunton Press, 2002. ***Concrete at Home***. Newton, CT: Taunton Press, 2005. *Concrete Countertops* was a surprising bestseller, dealing with design and techniques, with luscious photographs. *Concrete at Home* expands into floors, walls, columns and fireplaces.

Folkard, Edward. ***Casting in Ciment Fondu***. Alec Tiranti, 70 High Street, Theale, Reading, Berks, England, 1983.

Hunter, Sherri Warner. ***Creative Concrete Ornaments for the Garden: Making Pots, Planters, Birdbaths, Sculpture & More***. New York: Lark Books, 2005. Similar to *Making Concrete Garden Ornaments* but more complete. ***Making Concrete Garden Ornaments*** *(formerly: Creating With Concrete: Yard Art, Sculpture, and Garden Projects, 2001)*. New York: Lark Books, 2002. Excellent instructions for different types of concrete using many techniques: casting, armatures, foam, carving, surface treatments.

James, Thurston. ***The Prop Builder's Molding & Casting Handbook***. Cincinnati: Betterway Books, 1989. A good overview of mould-making techniques from an experienced prop builder. A lot of this information can be applied to concrete casting.

Kowal, Dennis and Meilach, Dona Z. ***Sculpture Casting: Mold Techniques and Materials: Metals, Plastics, Concrete***. New York: Crown Publishing, 1972. This book has a chapter on concrete and plaster casting, with other sections on mould making. A good book for large-scale work and plaster molds.

Majumdar, A.J. and Laws, V. ***Glass Fibre Reinforced Cement***. Oxford: BSP Professional Books, Oxford, 1991. This is a highly technical research book on the combination of alkali resistant glass fibres and cement. It may be difficult to find. I mention it as there are few books on this specialized subject.

Mills, John W. ***Sculpture in Concrete***. New York: Praeger, 1968. This book describes making a multi-piece plaster mould around a clay form. The mould can then be used to cast one solid concrete sculpture.

Mills, John W. ***The Technique of Sculpture***. London: B.T. Batsford Ltd., 1990.

Naaman, Antoine E. ***Ferrocement and Laminated Cementitious Composites***. Techno Press, 2000. Basic ferro-cement text frequently recommended on ferro-cement discussion lists.

Olson, Lynn. ***Sculpting With Cement: Direct Modeling in a Permanent Medium***. Valparaiso, Indiana: Steelstone Press, 1982. There are excellent sections on surface finishing, reinforcing with steel, carving and additives.

Plowman, John. ***The Encyclopedia of Sculpting Techniques***. Philadelphia, London: Running Press, 1995. Wide variety of techniques, none using concrete except for ciment fondu casting and carving aerated blocks. Sections on waste moulds, rubber moulds and casting, much of which could be adapted for concrete.

Portland Cement Association. ***Design and Control of Concrete Mixtures***. Published in Canada (www.cement.ca) and the United States (www.cement.org), these are excellent technical references with a new focus on supplementary materials, fibres and lightweight mixtures.

Schwanke. Dik and Lahti-Wagner, Jean. ***Cement Sculpture: A Studio Handbook***. Lanham, MD: University Press of America, 1985. This handbook describes the technique of carving into cement when it is partially set to give rapid and pleasing results.

Vitruvius. ***The Ten Books of Architecture***. New York. Dover Publications, 1960 (1914). Translated by Morris Hicky Morgan. Written by Marcus Vitruvius Pollio, Roman architect and engineer in the first century B.C., this book includes what are probably the first brief writings about concrete, lime and pozzolans. Historically interesting.

C

Links on the Internet

The following links can be accessed easily by going to my own web site's links page at: **http://www.makersgallery.com/concrete/links.html**. The online links will probably be more current than those listed below. I have divided the list under these headings:

- **GENERAL CONCRETE**
- **SPECIALIZED CONCRETE**
- **ART, ARTISTS, DESIGN, ARCHITECTURE**
- **SUPPLIERS, MANUFACTURERS**
- **OTHER CONCRETE ISSUES**

My intent is to keep this handbook up to date by posting information on the web site that started this process. The web page will refer the reader to new information such as updates for sources of supply, materials and corrections:
http://www.makersgallery.com/concrete/handbook.html

General Concrete

http://matse1.mse.uiuc.edu/~tw/concrete/prin.html
University of Illinois page has good solid scientific principles about concrete.

http://www.cement.org/tech/cct_port_cem_prod_tech.asp
Good overview from the Portland Cement Association.

http://matse1.mse.uiuc.edu/~tw/concrete/concrete.html
Concrete information: teaching materials, history timeline, scientific principles. University of Illinois, Materials Science and Engineering.

http://www.concretenetwork.com/
Concrete Network is a good overall site for house construction issues, pavers, countertops.

Specialized Concrete

http://www.litracon.hu/menu_en.htm
LiTraCon, light-transmitting concrete developed by Aron Losonczi a Hungarian inventor

http://ferrocement.net/mailman/listinfo/list_ferrocement.net
Ferro-cement email discussion list.

http://irc.nrc-cnrc.gc.ca/pubs/cbd/cbd103_e.html
Canadian Building Digest, good article on admixtures in concrete.

http://irc.nrc-cnrc.gc.ca/pubs/cbd/cbd241_e.html
Canadian Building Digest, why polymer modified concretes are recommended.

http://www.decorative-concrete.net/
Decorative-concrete.net has links to the decorative and stamped concrete industry.

http://www.elvesofester.com/mosaics.html
Cole Sonafrank's stained mosaic mortar. Discusses components of a good mix, also links to standard mix page and a good list of suppliers.

http://www.siliconfolly.com/
Good overview of process and materials for stained glass stones.

http://whatcom.wsu.edu/ag/homehort/hypertufa.htm
http://www.efildoog-nz.com/hypertufa.htm
Good hyper tufa introductions with instructions.

http://www.fhwa.dot.gov/infrastructure/materialsgrp/silica.htm
Silica Fume information from US Department of Transportation.

Art, Artists, Design, Architecture

http://cowboy-zen.com/
Grayson Malone, Kensington, California, artist working in metal infused concretes. Large gallery of images.

http://www.concrete-art.com/
Charles Sthreshley, Virginia artist, makes concrete sculpture and furniture. Excellent, extensive how-to sections on using steel armatures and applying the concrete.

http://www.faducci.com
Cement, steel and mosaic sculpture, bright colours, North San Juan, CA.

http://www.quicksandstudio.com
Martin Webb, coloured concrete wall pieces, San Francisco area.

http://www.midcoast.com/~bo
Bo Atkinson, works in concrete, pools, fountains, concepts.

http://www.perranch.com/
Contemporary geometric concrete from Sweden.

http://www.el-pasillo.com.ar/
Graciela Petroni is an architect from Buenos Aires, Argentina, working in concrete and iron furniture.

http://www.theconcretist.com
"An association of independent artists, craftsmen and others who specialize in producing sensory concrete art and architecture." International.

http://www.littleandlewis.com/
Concrete sculpture and paintings for the home and garden. Bainbridge Island, Washington.

http://www.sculptor1.com/
John Wilkinson, figurative sculptor in concrete and gypsum, Stonington, Maine.

http://www.gosculpture.com/
Jeff Yielding, abstract cement and wood sculpture.

http://www.chengdesign.com/
Custom homes, concrete countertops and NeoMix concrete pre-mix and supplies. Author of the book Concrete Countertops.

http://www.cement-creativity.org.uk/gallery.html
British Cement Association Awards for students to celebrate design and innovation using concrete.

http://www.concretedecisions.com
Preston Manser, industrial designer, makes architectural elements and objects for landscaping in concrete, Hilton Head Island, South Carolina.

http://www.oceanstonestudio.com/
Bara Fallows and Richard MacKenzie make concrete garden art, Galiano Island, BC.

http://www.lagniappemosaic.com/
Andree LeBlanc is Florida-based craftsperson making mosaic work.

http://www.ancientartifactstile.com
http://www.ilenedunn.com
Ilene Dunn creates work in concrete and mortar, including tiles.

http://www.theshedstudios.com/
Rachel Reynolds in London, England, shows contemporary garden furniture in concrete.

http://www.compactconcrete.com/index_en.html
Francesco Passaniti, Paris, designs furniture, objects and sculptures, many in concrete.

http://www.concrete-clay.de
Two artists from Berlin have this web site called concrete+clay. In German.

http://www.taconic.net/kanwit/method.htm
Roy Kanwit. Spencertown, New York, sculptor shows the process of making a 20-foot high head in concrete and steel mesh.

http://users.lmi.net/~drewid/Techniques2.htm
Andrew Werby shares sculpture techniques including concrete.

http://www.buddyrhodes.com/
Buddy Rhodes Studio, San Francisco. Minimal concrete furniture, floor tiles, etc.

http://garden-art.com/
Garden Art Gallery, Chapel Hill, NC, has clay, marble and some concrete.

http://www.suedorman.com
Sue Dorman at Loyola Marymount University has worked a lot with concrete in jewelry.

http://www.arcosanti.org/
Arcosanti, an experimental town in the high desert of Arizona, is built with concrete. They offer workshops where you can help build the community.

http://www.stonewear.com/
Stonewear makes GFRC (glass fibre reinforced concrete) fountains and columns.

http://www.betonskulpturen.ch
Jacqueline Scholten and Hannelore Holz in Switzerland make life-size concrete figures painted with acrylics.

http://www.voile-de-beton.com/
Thierry Bogaert, architect designer in Paris, France, creates Voile de Beton (the scent of concrete) in these concrete perfume bottles.

http://www.sandbackfab.com
Sandback Fabrication in New Hampshire makes simple geometric tables with coloured concrete.

Concrete Safety

http://www.msds.com/
MSDS Solutions. Search by name or manufacturer. See also:
http://www.setonresourcecenter.com/MSDS/index.htm

http://www.netic.com.ar/sculp/schaz.htm
Traditional Sculpture Hazards, a text site by Michael McCann, Ph.D., C.I.H., and Angela Babin, M.S., contains a small section on Portland cement. Lots of information on other sculpture materials.

Suppliers and Manufacturers

http://www.sculpt.com
The Compleat Sculptor supplies alkali-resistant fibre (chopped and mat), latex additives, pigments and a variety of aggregates. They also carry Winterstone® and rubber moulding compounds. New York, NY.

http://www.euclidchemical.com

In Canada and the USA, supplies all types of additives, fibres, silica fume, materials for concrete.

http://www.artcast.com/art-supplies.html
Artcast in Georgetown, ON, has casting supplies.

http://www.tiranti.co.uk/
Alec Tiranti, sculpture supply, United Kingdom.

http://www.ball-consulting-ltd.com/
Ball Consulting Ambridge, PA, USA, supplies alkali-resistant glass fibres and acrylic as cement additives. The site also shows Hiram P. Ball's sculpture.

http://www.fibermesh.com/
Fibermesh makes polypropylene fibre additives for concrete.

http://www.nycon.com
Nycon makes nylon fibres for concrete.

http://www.cem-fil.com/index.html
Cem-FIL International "promotes and sells alkali resistant glassfibres as a primary reinforcement for cement based matrices." A good information-filled site.

http://www.kuraray-am.com/pvaf/index.php
Kuraray America makes PVA (polyvinyl alcohol) fibers.

http://www.b-lite.com/bsttext.htm
B-Lite makes polystyrene beads coated to attract cement paste.

http://www.degussaadmixtures.com/
Master Builders, Inc. is involved in research and development and has an extensive site with information on mixtures, additives, finishes, and concrete canoes.

http://www.formandbuild.com
Form and Build is an Ontario, Canada, distributor of supplies to the construction industry. (Sells siloxane sealer, sealers, fibres.)

http://specco.com/
Manufacturers of a large selection of admixtures, retardants, accelerators, form releases, and sealers for concrete.

http://www.concretecanada.com/
Conroc Systems Ltd. is a Canadian supplier and contractor...using glass fiber reinforced concrete in Calgary, Alberta. Suppliers for decorative concrete, including acid stains, polymers and sealers.

http://www.arizonaoxides.com
Colors for Concrete (Arizona Oxides) has a good overview of pigment use.

http://www.solomoncolors.com/concreteclrs.htm
SGS Concrete Product Colors has a good colour chart to show the limited palette of pigments available.

http://www.kemiko.com

http://www.concrete-stains.com (Triple-S Chemical)
http://www.negeoproducts.com
Sources for acid stains for concrete.

http://www.fillite.com/
Fillite®: hollow alumino-silicate cenospheres or microspheres derived from fly ash, used for weight reduction.

http://www.engelhard.com/
Engelhard manufactures Metamax, their brand of metakaolin.
http://www.engelhard.com/resources/MetaMax/MetaMax/
An exceptional overview of pozzolans, in particular metakaolin (Metamax) and how cement works.

http://www.metakaolin.com/
Advanced Cement Technologies, excellent reference articles on metakaolin.
http://www.concretecountertopinstitute.com
The Concrete Countertop Institute sells individual bags (40 lbs) of metakaolin in the U.S.A.

http://www.winterstone.com/
Winterstone is a cement-based product made for sculptors. The site has excellent sections on casting and armature-type processes.

http://www.milestonehpc.com/
"Milestone Hybridized Portland Cement is a specially formulated acrylic admixture for use in Portland Cement." You can order and purchase the product.

http://www.polyadam.com/
A complete system, combining acrylic polymer additive, Portland cement, sand, and glass scrim. Scrim is fiberglass mesh, alkali resistant, which can be used instead of hardware cloth.

http://www.vseal.com
V-Seal manufactures an effective water-based sealer for concrete. Also available in Canada (Toronto) through simon@concreationcanada.com

http://www.seal-it.sk.ca/
Seal-It, a Saskatoon, Saskatchewan, company, makes a line of concrete sealers.

http://miraclesealants.com/sealers.html
Manufacturers of 511 Porous Plus, a penetrating sealer for concrete and stone that does not change the surface quality. Can be ordered online.

http://www.c-cure.com/products/special/main.htm
C-cure's Penetrating Sealer 978, water based, silicone type water repellent.

http://www.epoxyproducts.com/vee4u.html
Bio Vee, another concrete sealer that reacts with lime, is available here.

http://www.graceconstruction.com
Grace Construction is a supplier of catalysts, silica products, specialty construction chemicals, building materials and container sealants.

http://www.fox-ind.com

Fox Industries has developed several admixtures and has an online catalog (Maryland, USA).

http://www.grtinc.com/products/melchem.cfm
General Resource Technology makes Melchem, a water-soluble, melamine-based, water-reducing admixture (a superplasticizer).

http://www.sculpt.com/catalog_98/moldingaccessories/smoothon/in+out.htm
http://www.cresset.com/ra.htm
For mould release agents.

http://www.polytek.com
Supplier of mold-making equipment, rubber, resin and supplies, in Easton, PA. Online store.

http://www.cementics.com/
Milestone, a hybridized Portland cement, is a specially formulated acrylic admixture for art and sculptural use.

http://www.artstuf.com/
Douglas and Sturgess Art Supplies, San Francisco, CA. Mold release agents, metal powders, glass spheres for lightweight concrete.

http://www.cpd.ca
CPD, Construction Product Distribution Services in Concord, Ontario, supplies concrete-related products.

Other Concrete Issues

http://www.umanitoba.ca/cast_building/
Centre for Architectural Structures and Technology at the University of Manitoba has been doing interesting research into flexible fabric formworks.

http://irc.nrc-cnrc.gc.ca/pubs/cbd/cbd215_e.html
Waste and By-Products as Concrete Aggregates: 1981 paper from National Research Council, Canada.

http://www.sbtc.ca/
Ecosmart Concrete Project, a web site about more environmentally friendly concrete.

http://www.romanconcrete.com/
Romanconcrete.com: concrete from 1,800 years ago.

http://www.civil.columbia.edu/meyer/glascrete.html
Good summary of concrete research at Columbia University using scrap glass.

http://www.geocities.com/flyingconcrete/Other/links.htm
Useful links to concrete sites including sculpture.

d

MSDS Portland Cement

This Material Safety Data Sheet (MSDS) has been compiled from several manufacturers' Material Safety Data Sheets. *It is not complete and cements can differ in their makeup.* You should obtain the latest MSDS from the supplier of the Portland cement you purchase along with the MSDS of any other hazardous material. I have included it here to emphasize the safety precautions you should be taking when working in concrete, and as an example of what is easily available for any hazardous materials you may be working with.

Chemical Family:
Calcium compounds. Calcium silicate compounds and other calcium compounds containing iron and aluminum make up the majority of this product.

Chemical Name And Synonyms:
Portland cement. Portland cement is also known as hydraulic cement and/or normal Portland cement.

Formula:
This product consists of finely ground Portland cement clinker, gypsum and limestone (for some products).

Trace Ingredients:
Due to the use of substances mined from the earth's crust, trace amounts of naturally occurring, potentially
uents may be found in Portland cement. Portland cement may contain up to 0.75% insoluble
ll amount of this residue includes free crystalline silica. Portland cement also may contain trace
unts of chromium salts or compounds (including hexavalent chromium) or other metals (including
nds) found to be hazardous or toxic in some chemical forms. These metals are present mostly as
ions within the principal minerals. Other trace constituents may include potassium and sodium
unds.

AND FIRST AID

INFORMATION:
ent is a light gray or white powder. When in contact with moisture in eyes or on skin, or when
ater, Portland cement becomes highly caustic (pH > 12) and will damage or burn (as severely as third-degree) the eyes or skin. Inhalation may cause irritation to the moist mucous membranes of the nose, throat and upper respiratory system or may cause or may aggravate certain lung diseases or conditions. Use exposure controls or personal protection methods described in the section on Stability and Reactivity.

EYES:
Immediately flush eye thoroughly with water. Continue flushing eye for at least 15 minutes, including under lids, to remove all particles. Call physician immediately.

SKIN:
Wash skin with cool water and pH-neutral soap or a mild detergent. Seek medical treatment if irritation or inflammation develops or persists. Seek immediate medical treatment in the event of burns.

INHALATION:
Remove person to fresh air. If breathing is difficult, administer oxygen. If not breathing, give artificial respiration. Seek medical help if coughing and other symptoms do not subside. Inhalation of large amounts of Portland cement requires immediate medical attention.

INGESTION:
Do not induce vomiting. If conscious, have the victim drink plenty of water and call a physician immediately.

HAZARDS IDENTIFICATION/TOXICOLOGICAL INFORMATION

Emergency Overview:
Portland cement is a light gray powder that poses little immediate hazard. A single short-term exposure to the dry powder is not likely to cause serious harm. However, exposure of sufficient duration to wet Portland cement can cause serious, potentially irreversible tissue (including skin or eye) destruction in the form of chemical (caustic) burns, including third degree burns. The same type of tissue destruction can occur if wet or moist areas of the body are exposed for sufficient duration to dry Portland cement.

Potential Health Effects:
When mixed with water, a strong alkaline solution is produced. If this solution comes in contact with eyes or skin it may cause serious burns and ulceration. The eyes are particularly vulnerable and damage will increase with contact time. Strong alkaline solutions in contact with skin tend to damage the nerve endings first before damaging the skin, therefore chemical burns can develop without pain being felt at the time.

Potential effects resulting from eye contact:
Exposure to airborne dust may cause immediate or delayed irritation or inflammation. Eye contact by larger amounts of dry powder or splashes of wet Portland cement may cause effects ranging from moderate eye irritation to chemical burns and blindness. Such exposures require immediate first aid (see Section 4) and medical attention to prevent significant damage to the eye...Soft contact lenses may absorb irritants and all lenses concentrate them.

Potential effects resulting from skin contact:
Discomfort or pain cannot be relied upon to alert a person to a hazardous skin exposure. Consequently, the only effective means of avoiding skin injury or illness involves minimizing or avoiding skin contact, particularly contact with wet cement. Persons exposed to wet cement may not feel discomfort until hours after the exposure has ended and significant injury has occurred.

Exposure to dry Portland cement may cause drying of the skin with consequent mild irritation or more significant effects attributable to aggravation of other conditions. Dry Portland cement contacting wet skin or exposure to moist or wet Portland cement may cause more severe skin effects including thickening, cracking, or fissuring of the skin. Prolonged exposure can cause severe skin damage in the form of (caustic) chemical burns.

Some individuals may exhibit an allergic response upon exposure to Portland cement, possibly due to trace amounts of hexavalent chromium. The response may appear in a variety of forms ranging from a mild rash to severe skin ulcers. Persons already sensitized may react to their first contact with the product. Other persons may experience this effect after years of contact with hydraulic cement products.

Potential effects resulting from inhalation:
Portland cement may contain trace amounts (<0.1%) of free crystalline silica. Prolonged exposure to respirable free crystalline silica may aggravate other lung conditions. It may also cause delayed lung injury including silicosis, a disabling and potentially fatal lung disease and/or other diseases.

Exposure to Portland cement may cause irritation to the moist mucous membranes of the nose, throat, and upper respiratory system. It may also leave unpleasant deposits in the nose.

Potential effects resulting from ingestion:
Although small quantities of dust are not known to be harmful, ill effects are possible if larger quantities are consumed. Portland cement should not be eaten under any circumstances.

Carcinogenic Potential:
Portland cement is not listed as a carcinogen by IARC, NTP, or OSHA. It may, however, contain trace amounts (<0.1%) of substances listed as carcinogens by these organizations. Crystalline silica is now classified by IARC as a known human carcinogen (Group I). NTP had characterized respirable crystalline silica as "reasonably anticipated to be (a) carcinogen."

Medical Conditions Which May Be Aggravated By Inhalation or Dermal Exposure:
- Pre-existing upper respiratory and lung diseases.
- Unusual (hyper) sensitivity to hexavalent chromium (chromium +6) salts.

FIRST AID

Eyes:
Immediately flush eyes thoroughly with water. Continue flushing eye for at least 15 minutes, including under lids, to remove all particles. Call physician immediately.

Skin:
Wash skin with cool water and pH-neutral soap or a mild detergent intended for use on skin. Seek medical treatment in all cases of prolonged exposure to wet cement, cement mixtures, liquids from fresh cement products, or prolonged wet skin exposure to dry cement.

Inhalation of Airborne Dust:
Remove to fresh air. Seek medical help if coughing and other symptoms do not subside. (Inhalation of gross amounts of Portland cement requires immediate medical attention.)

Ingestion:
Do not induce vomiting. If conscious, have the victim drink plenty of water and call a physician immediately.

FIRE & EXPLOSION DATA

Flammability: Not Flammable.
Flash Point: Not Applicable.
Lower Explosive Limit: Not Applicable.
Upper Explosive Limit: Not Applicable.
Auto ignition Temperature: Not Applicable.
Sensitivity to Static Discharge: Not Applicable.
Sensitivity to Impact: Not Applicable.
Extinguishing Media: Not Applicable.
Special Fire-Fighting Procedures: None.
Hazardous Combustion Products: Not Applicable.
Unusual Fire and Explosion Hazards: Not Applicable.

ACCIDENTAL RELEASE MEASURES

Collect dry material using a scoop. Avoid actions that cause dust to become airborne. Avoid inhalation of dust and contact with skin. Wear appropriate personal protective equipment as described in the section on Exposure Control/Personal protection.
Scrape up wet material and place in an appropriate container. Allow the material to "dry" before disposal. Do not attempt to wash Portland cement down drains.
Dispose of waste material according to local, state, and federal regulations.

HANDLING AND STORAGE

Keep Portland cement dry until used. Normal temperature and pressure do not affect the material.

Promptly remove dusty clothing or clothing which is wet with cement fluids and launder before reuse. Wash thoroughly after exposure to dust or wet cement mixtures or fluids.

EXPOSURE CONTROLS / PERSONAL PROTECTION

Skin Protection:
Prevention is essential to avoiding potentially severe skin injury. Avoid contact with unhardened (wet) Portland cement products. If contact occurs, promptly wash affected area with soap and water. Where prolonged exposure to unhardened Portland cement products might occur, wear impervious clothing and gloves to eliminate skin contact.

Respiratory Protection:
Avoid actions that cause dust to become airborne. Use local or general ventilation to control exposures below applicable exposure limits. Use NIOSH/MSHA-approved respirators in poorly ventilated areas when dust causes discomfort or irritation, or where there is an applicable exposure limit (Advisory: Respirators and filters purchased after July 10, 1998 must be certified under 42 CFR 84).

Ventilation:
Use local exhaust or general dilution ventilation to control exposure within applicable limits.

Eye Protection:
When engaged in activities where cement dust or wet cement or concrete could contract the eye, wear safety glasses with side shields or goggles. In extremely dusty environments and unpredictable environments, wear unvented or indirectly vented goggles to avoid eye irritation or injury. Contact lenses should not be worn when working with Portland cement or fresh cement products.

PHYSICAL AND CHEMICAL PROPERTIES

Appearance: White to gray powder.
Odor: No distinct odor.
Odor Threshold: Not applicable.
Physical State: Solid (powder).
pH (as a solid): Not applicable.
pH (in water) (ASTM D 1293-95): 12 to 13
Solubility in Water: Slightly soluble (0.1 to 1.0%)
Vapor Pressure: Not applicable.
Vapor Density: Not applicable.
Boiling Point: Not applicable (i.e.,>1000oC).
Freezing Point: Not applicable.
Melting Point: Not applicable.
Specific Gravity (H20 =1.0):3.15.
Evaporation Rate: Not applicable.
Coeff. Water/Oil Dist.: Not applicable.

STABILITY AND REACTIVITY

Stability: Stable
Conditions to avoid: Unintentional contact with water.
Incompatibility: Portland cement reacts with water to produce a caustic solution, pH 12 to pH 13. Wet Portland cement is alkaline. As such it is incompatible with acids, ammonium salts and aluminum metal. Aluminum powder and other alkali and alkaline earth elements will react in wet mortar or concrete, liberating hydrogen gas. Portland cement dissolves in hydrofluoric acid producing corrosive silicon tetraflouride gas. Silicates react with powerful oxidizers such as fluorine, chlorine, trifluoride and oxygen difluoride
Hazardous Decomposition: Will not spontaneously occur. Adding water results in hydration and produces (caustic) calcium hydroxide. Thermal decomposition may increase calcium oxide content.
Hazardous Polymerization: Will not occur.

TOXICOLOGICAL INFORMATION

Effects Of Acute Exposure:
Portland cement and wet Portland cement mixtures can dry the skin, cause alkali burns and irritate the eyes and upper respiratory tract. Ingestion can cause irritation of the throat.

Effects Of Chronic Exposure:
Portland cement dust can cause inflammation of the tissue lining the interior of the nose and the cornea (white) of the eye.

ECOLOGICAL INFORMATION

Ecotoxicity: No recognized unusual toxicity to plants or animals.
Relevant Physical and Chemical Properties: see sections on Physical and Chemical Properties and Stability and Reactivity.
Dispose of waste material according to local, provincial, state and federal regulations. (Since Portland cement is stable, uncontaminated material may be saved for future use.)
Dispose of bags in an approved landfill or incinerator.

TRANSPORT INFORMATION

Hazardous materials description/proper shipping name: Portland cement is not hazardous under the TDG Act (Canada) or DOT regulations (USA).
Hazard Class: Not applicable.
Identification Number: Not applicable.
Required Label Text: Not applicable.
Hazardous substances/reportable quantities (RO): Not applicable.

STATUS

Status under USDOL-OSHA Hazard Communication Rule, 29 CFR 1910.1200:
Portland cement is considered a "hazardous chemical" under this regulation, and should be part of any hazard communication program.

Status under CERCLA/Superfund, 40 CFR 117 and 302: Not listed.

Hazard Category under SARA (Title III), Sections 311 and 312:
Portland cement qualifies as a "hazardous substance" with delayed health effects.

Status under SARA (Title III), Section 313: Not subject to reporting requirements under Section 313.

Status under TSCA (as of May 1997): Some substances in Portland cement are on the TCSA inventory list.

Status under the Federal Hazardous Substances Act:
Portland cement is a "hazardous substance" subject to statutes promulgated under the subject act.

Status under California Proposition 65:
This product contains chemicals (trace metals) known to the State of California to cause cancer, birth defects or other reproductive harm. California law requires the manufacturer to give the above warning in the absence of definitive testing to prove the defined risks do not exist.

Status under Canadian Environmental Protection Act: Not listed.

Status under WHMIS:
Portland cement is considered to be a hazardous material under the Hazardous Products Act as defined by the Controlled Products Regulations and is therefore subject to the labeling and MSDS requirements of the Workplace Hazardous Materials Information System (WHMIS).

Portland cement has been classified in accordance with the hazard criteria of the CPR and the MSDS contains all the information required by the CPR.

OTHER INFORMATION

Portland cement should only be used by knowledgeable persons. A key to using the product safely requires the user to recognize that Portland cement chemically reacts with water, and that some of the intermediate products of this reaction (that is, those present while a Portland cement product is "setting") pose a far more severe hazard than does Portland cement itself.

While the information provided in this material safety data sheet compilation is believed to provide a useful summary of the hazards of Portland cement as it is commonly used, the compilation cannot anticipate and provide all of the information that might be needed in every situation. Inexperienced product users should obtain proper training before using this product.

In particular, the data furnished in this sheet does not address hazards that may be posed by other materials mixed with Portland cement to produce Portland cement products. Users should review other relevant material safety data sheets before working with Portland cement or working on Portland cement products.

No representations or warranties with respect to the accuracy or correctness of this information, or of any kind or nature whatsoever are given, made or intended by the author or the publisher. No legal responsibility whatsoever is assumed for this information, or for any injuries or damages, however caused which may result from the use of this information. This information is offered solely for informational purposes and is subject to your own independent investigation and verification.

e

Glossary

Accelerator: a material added to concrete to shorten the set time by increasing the rate of hydration.

Admixture: a material (such as a plasticizer) added to concrete in smaller quantities. Sometimes it is referred to as an **admix** or **additive**.

Aerated concrete: lightweight, carveable concrete with a uniform cellular structure (like a sponge), usually made from a mixture of Portland cement, sand, and a gas-forming agent such as aluminum, or an introduced foam. Commercial aerated concrete is steam-cured under pressure.

Air entrainer: an admixture that traps tiny bubbles of air in the concrete mix to improve frost resistance.

Alkaline: describes a material with a pH of over 7. Alkaline materials are also described as basic or caustic. Alkaline materials can neutralize acidic ones.

Armature: a metal framework on which concrete is placed to make a sculpture or other object.

Blast furnace slag: Ground, granulated blast furnace slag (GGBFS) is a recycled product. One of several *pozzolans* that can be added to concrete.

Bronchitis: inflammation of the mucous membrane in the bronchial tubes.

Calcium hydroxide: one of the two main products resulting from the hydration of Portland cement. It is commonly known as slaked lime and can burn the skin. The other is calcium silicate hydrate.

Calcium silicate hydrate: the main binding material in concrete, resulting from the hydration of Portland cement. See *calcium hydroxide*.

Caustic: capable of burning organic tissue; strongly alkaline.

Cement: any substance used to make things stick together. In the context of this handbook cement is Portland cement, defined as the fine grey powder that is made by burning limestone and clay. Calcium silicates are the active part of cement; they chemically combine with water to bind all the aggregates together.

Cement paste: a mixture of cement and water.

Cementitious: having cement-like properties. Pozzolans react with calcium hydroxide in concrete to make cementitious materials.

Ciment fondu: also known as calcium aluminate cement, or high alumina cement. A very fast setting cement originally developed for its sulphate resistance. It is chemically different than Portland cement.

Concrete: a mixture of crushed stone, sand, cement and water that hardens as a result of chemical reactions between the cement and water. "Concrete" means the wet mixture and also the hard material that results when it sets.

Curing: the process of letting concrete develop strength by keeping it damp and within the correct temperature range, usually for at least a week.

ECC: “engineered cementitious composites” is a term used to describe new sophisticated concrete mixes using such additives as pozzolans or fibres.

Emphysema: an enlargement of the air sacs of the lungs causing breathlessness.

Expanded lath or metal: galvanized steel that has been slit in a regular pattern and stretched; used as reinforcement for stucco.

False set: an initial rapid hardening of concrete, not caused by heat or accelerators. The concrete can be remixed and used. Not to be confused with a true *initial set*.

Ferro-cement or **ferroconcrete**: a form of reinforced concrete using closely spaced multiple layers of steel mesh or small rods completely encased in a cement-sand mix.

Final set: an arbitrary point at which concrete has hardened or become rigid. (See *initial set*.)

Fly ash: a fine-grained glassy powder collected from the flue gases of coal-burning power plants. One of several *pozzolans* that can be added to concrete.

Form (in the context of this book): the shape of something, apart from its colour and material.

Hardware cloth: a heavy-duty screening made from galvanized steel wire.

Hydration: in concrete, the chemical reaction between Portland cement and water. The longer concrete is kept damp, the longer the hydration will occur and the stronger the concrete will be.

Initial set: the first stiffening or hardening that takes place after concrete is mixed. This usually takes place in the first four to eight hours. (See *false set* and *final set*.)

Lime: calcium oxide. In cement the term usually means slaked lime (calcium hydroxide), a caustic substance. Lime is often an ingredient in mortar.

Limestone: a sedimentary rock consisting mainly of calcium carbonate.

Marble: a form of limestone (calcium carbonate) in a metamorphic crystalline state.

Metakaolin: a white kaolin clay that has been heat-treated. One of several *pozzolans* that can be added to concrete to increase strength and workability.

Mortar: a mixture of cement, sand and water. Mortar that is used for laying bricks or stones sometimes has added lime.

Mould (mold): a hollow shape in which something is cast. Moulds can be made of any material, the common ones being metal, wood and plaster.

Perlite: a type of volcanic glass that is expanded by heating, and used as a planting medium, an insulation, and to make lightweight concrete.

Permeability: the ability of a material to allow gas or water vapour to pass through it. Denser concrete has a higher impermeability.

pH: a chemical value measuring the acidity or alkalinity of a substance. A pH of 7 is neutral. Under 7 is acidic; over 7 is alkaline. Wet Portland cement has a pH of 10 to 12. (See *caustic*.)

Plasticizer (superplasticizer**):** an admixture that increases the workability of concrete, allowing less water in the mix. Sometimes referred to as a water reducer.

Polymer concrete: a concrete in which aggregates and minerals are held together by polymer resins, usually polyester, acrylic or epoxy, replacing some or all of the Portland cement in the mix. This is done to increase moisture resistance and strength.

Polystyrene: a rigid expanded plastic foam that can be used as a basic shape over which an armature and concrete is applied.

Polystyrene beads: lightweight foam beads that can be used as a lightweight aggregate to lighten a concrete mix.

Pozzolans: silicates (powders such as fly ash, or kaolin) that are sometimes added to cement to improve its properties. The silicates combine with the calcium hydroxide in the cement paste to produce insoluble *cementitious* substances. *Pozzolana* was a volcanic ash found near Pozzuoli and used in Roman times to make cement by mixing it with lime.

Premix: bagged dry mix to which you just add water. Not to be confused with *ready-mix*.

PVA: or polyvinyl alcohol fibres are a recently developed fibre additive in ECCs. These plastic fibres make a molecular bond with concrete (unlike polyester or nylon fibres), adding tensile strength.

Ready-mix: concrete ordered from a ready-mix supplier (by the cubic meter or yard) delivered to a job site by truck. Not to be confused with *premix*.

Rebar: steel reinforcing rod used in concrete.

Reinforced concrete: concrete that has been strengthened by embedding a reinforcing material in it, usually steel.

Retarder (retardant): an admixture that extends the setting time of concrete.

Set: a term used to describe when concrete becomes hard or rigid.

Shrinkage cracks: Cracks due to shrinkage or temperature as the volume of concrete contracts usually within the first few days of placement.

Silica fume: a silicate, a by-product of producing silicon metal or ferrosilicon alloys. It is a very fine dark grey powder. One of several *pozzolans* that can be added to concrete.

Silicosis: a lung disease caused by the inhalation of dust containing silica.

Stucco: a type of mortar that is applied to the outside of buildings to give a hard finish.

Vermiculite: A naturally occurring soft mineral that resembles mica; when heated it produces a lightweight, sterile, laminated structure. Used for plant soils, as an insulating material, and to make lightweight concrete.

Workability: the ease with which concrete can be mixed, put in place and finished without separation.

Metric conversions:
1" = 2.54 cm = 25.4 mm
1/16" = .0625" = 1.588 mm
1 mm = .039"
1 cm = .394"
1 litre = 1000 ml = 1.76 pints
1 fluid oz. = 1/20 imperial pint = 28.41 ml
1 fluid oz. = 1/16 US pint = 29.57 ml
1 imperial pint = 568.26 ml
1 US pint = 473.18 ml
1 kilogram = 2.205 pounds
1 gram (g) = .035 ounces
1 ounce = 28.35 grams

Index

A

accelerator, 27, 127
acid stains, 33, 85
  supplies, 111
acrylic admix, 28
acrylic sealer, 33
acrylic suppliers, 109
adding wet, 83
additives. *See* admixtures
admix
  acrylic, 28
  latex, 28
admix suppliers, 109
admixture, 127
Aerated concrete, 127
aggregates, 25–28
  coarse, 23
  fine, 23
air bubbles, 104
air entrainer, 127
air entrainer suppliers, 109
air entrainers, 26
alkali resistant glass fibre, 31
alkaline, 127
allergic dermatitis, 22, 26
application tools, 38
ARGF, 31
armature, 55, 127
armatures, jewellery, 62
artists, 89–99
Aspdin, Joseph, 9
ASTM standards, 20
Aviña, Joaquin, 98

B

binding wire, 29, 30, 109
blast furnace slag, 26, 127
blended cements, 21
Books, 112–13
bowls, 77
brick sand, 23
bronchitis, 127
bubbles, 84

C

calcium aluminate cement, 87
calcium hydroxide, 11, 127
calcium silicate hydrate, 127
carbon fibres, 31
carnauba wax, 34
carving tools, 38, 39
carving wet, 83
cast mix, 43
Casting Simple Shapes, 75–82
casting, foamboard, 76
caustic, 127
cavities, 84
cement, 7, 20–22, 127
  blended, 21
  paste, 7
  safety, 22
  warning, 22
  white, 21
cement fondu. *See* ciment fondu
cement suppliers, 108
cementitious, 128
chemistry, 10
chicken wire, 30
Ciment Fondu, 87–88, 128
clay, 82
CO_2, 101
colour, 32–33
  amount, 32
Colour and Finish, 32–34
compressive forces, 53
concrete, 7, 128
  definition, 7, 128
  premix, 35
  types, 20
connections, 74
cracks, 104
CSA standards, 20
curing, 50, 128

D

Definitions, 7–8
Dunn, Ilene, 95
dust, 17
dyes, 32–33
  supplies, 111

E

ECC, 128
emphysema, 128
Envronmental Concerns, 101–2
errors, 103
expanded lath, 30, 128
eye damage, 22
eye first aid, 22
eye irritation, 18, 122

F

false set, 128
ferro-cement, 8, 93, 128
ferroconcrete, 8, 128
fiberglass mesh, 31
FiberMesh, 30
fibre, 30
fibre suppliers, 109
final set, 128
finishing, 83–86
first aid, 22, 121
flaking, 105
fly ash, 21, 26, 102, 128
foamboard, casting, 76
force diagrams, 53
form, 128
frost resistance, 26
furniture, 98

G

Gallery, 89–99
glass, 23
glass fibre, 31
glass fibre mix, 47
glass fibre suppliers, 109
Glossary, 126–30
greenhouse gas, 101

H

HAC, 87
hairline cracks, 59, 104
hardening, 49
hardware cloth, 30, 128
hazards, 17, 18, 22
Hazards of Working in
  Concrete, 17–18
help, 103
high alumina cement, 87
Hogbin, Stephen, 89
holes, 84
How to - Introduction,
  13–15
hydration, 9, 128
hyper tufa, 45, 96

I

initial set, 49, 128
Isle of Portland, 10

J

jewellery methods, 58–64

K

kaolin, 129

L

lamp, 99
latex admix, 28
latex suppliers, 109
lath, 109
lightweight aggregates, 24
lightweight mix, 46
lime, 11, 128
limestone, 128
limestone dust, 25
Loney, Bill, 96

M

marble, 128
marble dust, 25
mask, 17
Material Safety Data
  Sheets, 17
Materials: What to Use, 19
mesh
  chicken wire, 30
  expanded steel, 30
  galvanized, 30
  lath, 30
Mesh - Step by Step,
  65–74
mesh mix, 42
mesh suppliers, 109
metakaolin, 26, 129
metakaolin suppliers, 110
metal filings, 25
metal tools, 39
Methods, 57
metric conversion, 42
metric conversions, 130
mistakes, 103
mix
  glass fibre, 47
  jewellery, 59
  lightweight, 46
  perlite, 46
  small scale, 59
  smooth, 44
  vermiculite, 46
Mixes, 41–47
mixing rules, 41
mixing tools, 37
modelling clay, 82
mold. *See* mould
mortar, 7, 129
mortar premix, 35, 36
mortar sand, 23
mould, 129
  jewellery, 61
  miscellaneous, 82
  sand, 82
  supplies, 110
  waste, 76
  wood and metal, 80
mould release, 28
MSDS, 17
MSDS Portland Cement,
  121–26

N

nylon fibres, 30

P

paint, 86
Pantheon, 9
paste, 7, 127
peat moss. *See* hyper tufa
perlite, 24, 129
perlite mix, 46
permeability, 129
pH, 129
pigments, 32
  supplies, 111
plastic fibres, 30
plastic sheet, 40
plasticizer, 129
plasticizer suppliers, 109
plasticizers, 27
Polymer concrete, 129
polypropylene fibres, 30
Polystyrene, 129
Portland cement, 9, 20
potassium sealer, 34
pottery clay, 82
pozzolans, 25, 102, 129
Premix, 35
pre-mix
  definition, 129

problems, 103
PVA, 30, 42, 59, 109, 118, 129

R

ready-mix, 129
rebar, 29, 129
Rees, Nicholas, 93
reinforced concrete, 7, 129
Reinforcement, 29–31
reinforcing, 55
reliefs, 95
retardants, 27
retarder, 129
Reynolds, Rachel, 91

S

safety
  supplies, 111
safety equipment, 18
sand mould, 82
sand premix, 35
sealer application, 86
sealers, 33
  acrylic, 33
  jewellery, 63
  potassium, 34
  silane, 33
  siloxane, 33
set, 130
Setting (Hardening) & Curing, 49–51
sharp sand, 23
shear forces, 53
shrinkage, 104
shrinkage cracks, 12, 130
silane sealer, 33
silica fume, 26, 130
silica fume suppliers, 110
silicosis, 22, 130
siloxane application, 86
siloxane sealer, 33
skin irritation, 22
slag, blast furnace, 26
small scale mix, 42
smooth mix, 44
Sources of Supply, 108–11
stainless steel fibres, 31
stains
  acid, 33, 85
  acrylic, 33
  surface, 33
steel mesh, 30
steel reinforcing, 29
steel suppliers, 109
step by step, mesh, 65
stone dust, 25
storage, 15
stress diagrams, 53
Stresses and Strength, 53–56
stucco, 130
stucco lath, 30
styrofoam beads, 24
superplasticizer suppliers, 109
superplasticizers, 27
suppliers, 108
supplies, 40, 108
surface cracks, 104
Surface Finishing, 83–86

T

Technical - Concrete & Cement, 9–12
temperature, 49
tensile forces, 53
thixotropic, 14
tiles, 95
tools, 37–40
  application, 38
  carving, 38
  metal, 39
  mixing, 37
Troubleshooting, 103–5
tufa. *See* hyper tufa

V

vermiculite, 24, 130
vermiculite mix, 46
Very Small Scale, 58–64
vibration, 104
V-seal, 34

W

walled forms, jewellery, 60
warnings, 17
waste mould, 76
water
  final strength, 13, 27
  quantity, 11, 14, 50
  temperature, 49
water glass, 34
wax, 34
white Portland cement, 21
wire, 29
workability, 130

About the Author

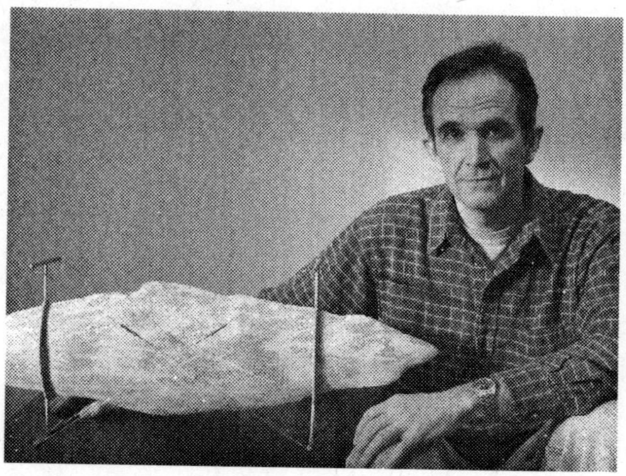

A professional craftsman, Andrew Goss has gained a reputation as an experimental and innovative designer and maker of jewellery. He studied at George Brown College in Toronto, Canada, Hornsey College of Art in London England, and the University of Toronto. Through explorations into non-precious jewellery materials such as paper and plastic, Goss was eventually led to concrete as a medium. He began creating, experimenting, and researching concrete in 1992, which led to making the first of his concrete and metal sculptures. Currently he now creates both jewellery and sculpture in the medium.

Since establishing Goss Design Studio in 1973, he has shown his work in over 100 exhibitions, including his show of concrete sculpture and jewellery at Prime Gallery in Toronto in 1999 and at Karsh-Masson Gallery in Ottawa in 2005. His work has been presented in galleries across Canada and the United States and is in several collections, including the Canadian Embassy in Berlin, Germany. He is a member of the Royal Canadian Academy of Arts.

Goss shares an Owen Sound, Ontario, Canada studio with his wife, jeweller Sandra Noble Goss.